GO WHERE FE[...] P9-BYM-474
BEEN ALLOWED BEFORE . . .
BIG RED

"Douglas Waller has done it again! In this real-life
version of *Crimson Tide*, America's best military
journalist takes us on an unforgettable patrol to
meet the men responsible for the U.S. Navy's
ultimate deterrent force. Packed with both personal
and technical details, if you want to learn any more
about the world's most awesome weapons platform,
you'll have to join the Navy!"
Richard P. Henrick, author of *Crimson Tide*

"Excellent . . . a richly-drawn account
that is eminently readable."
Undersea Warfare Magazine

"An eye-opening look into the top-secret world of
nuclear submarines. . . . Informative and readable."
Library Journal

"Submariners are a special breed. Because they serve
their country in a shroud of secrecy, they often fail
to get the recognition they deserve. *Big Red* gives us an
astonishing peek inside the daily life of a submarine
and a portrait of the men who make heartbreaking
personal sacrifices in order to serve their country.
Reading it, one can only develop awe and admiration
for the men of the 'Silent Service.'"
Jonathan Mostow, director and screenwriter of *U-571*

Also by Douglas C. Waller

The Commandos:
The Inside Story of America's Secret Soldiers
Air Warriors:
The Inside Story of the Making of a Navy Pilot

BIG RED

THE THREE-MONTH VOYAGE OF A TRIDENT NUCLEAR SUBMARINE

DOUGLAS C. WALLER

HarperTorch

An Imprint of HarperCollins*Publishers*

HARPERTORCH
An Imprint of HarperCollins*Publishers*
10 East 53rd Street
New York, New York 10022-5299

The HarperCollins hardcover edition contains the following Library of Congress Cataloging in Publication Data:

    Waller, Douglas C.
       Big Red : three months on board a Trident nuclear
    submarine /
    by Douglas C. Waller.—1st Harper trade ed.
       p.   cm.
    Includes index.
    1. Nebraska (Nuclear submarine).   2. Nuclear
    weapons—United States.   I. Title.
    VA65.N35 W35 2001
    359.9'330973—dc21   00-040880

First HarperTorch paperback printing: April 2002
First HarperCollins hardcover printing: March 2001

Visit HarperTorch on the World Wide Web at www.harpercollins.com

10   9   8   7   6   5   4   3

To Marjorie

• • •

There is no place like Nebraska
Dear old Nebraska
Where the girls are the fairest
The boys are the squarest
Of any a schools I knew

There is no place like Nebraska
Where we are all true blue
We'll all stick together
Through all kinds of weather
For dear old Nebraska

—NEBRASKA FIGHT SONG

• Contents

PART II: THE NEXT TWO MONTHS

PART III: GOING HOME

Nebraska Submariners Cited in This Book

Officers
Lt. Cdr. Duanc Ashton, former executive officer
Lt. Cdr. Alan Boyd, executive officer
Lt. Al Brady, navigator
Lt. J.G. Dave Bush, main propulsion assistant
Ens. Ray Chesney, bull ensign
Lt. Joe Davis, supply officer
Lt. Fred Freeland, weapons officer
Lt. Cdr. Harry Ganteaume, engineering officer
Lt. Steve Habermas, communications officer and Catholic
  lay leader
Lt. Ryan Hardee, nuclear weapons officer
Lt. J.G. Rob Hill, reactor division officer
Cpt. Jerry Hunnicutt, deputy squadron commander
Lt. Brent Kinman, drill coordinator
Ens. Mark Nowalk, junior ensign
Lt. J.G. Chad Thorson, damage control assistant
Lt. J.G. Bob Tremayne, tactical systems officer
Cdr. Dave Volonino, ship's captain

Chiefs
Marvin Abercrombie, departing A-ganger chief
Shawn Brown, assistant drill coordinator

Marc Churchwell, master chief from the squadron
Steve Dille, senior engineering chief
Stacy Hines, missile chief and watch chief
Bob Lewis, new chief of the A-gangers
Rich McCloud, mess chief
Sean McCue, missile chief
Dan Montgomery, communications chief
Bob Philbin, medical corpsman
Todd Snyder, sonar chief
Jeff Spooner, weapons chief and diving officer
Tom Standley, senior navigation chief
Dave Weller, chief of the boat

Petty Officers

Quentin Albea, yeoman and planesman
Jason Barrass, sonarman
Jason Bradfield, sonarman
Jason Bush, fire control technician and planesman
Eric Chambers, sonarman and lookout
Chris Cornelius, navigation petty officer
Paul Dichiara, missile technician
Jason Dillon, storekeeper
Matt Douvres, sonarman
Jason Duff, cook
Ben Dykes, sonarman
Ashley Fuqua, fire control technician and lookout
Richard Garvin, fire control technician
Tony Holmes, senior electrician
Tom Horner, radio operator
Calvin Ireland, missile technician
Kevin Jany, missile technician

Todd Jenkins, throttleman
James Johnson, sonarman
Don Katherman, missile technician
Keith Larson, torpedoman and planesman
Jason Lawson, missile technician
Don Lee, A-ganger
Frank Levering, Jack of the Dust
Eric Liebrich, radio operator
Mark Lyman, missile technician
Rodney Mackey, contact plotter and Protestant lay leader
Seth Magrath, A-ganger
Ed Martin, reactor petty officer, minister, and barber
Greg Migliore, radio operator
Greg Murphy, missile technician
Shawn Olmstead, electrician and barber
James Penn, storekeeper and watch chief
Reginald Rose, storekeeper and planesman
Brett Segura, machinery petty officer and watch chief
David Smith, torpedoman
Ed Stammer, navigation petty officer
Matt Suzor, navigation petty officer
Gregory Turner, machinist
Jose Victoria, missile technician
Chris Wilhoite, sonarman
Keith Williams, missile technician

New Seamen
Ryan Beasley
Miguel Morales
Scott Shafer

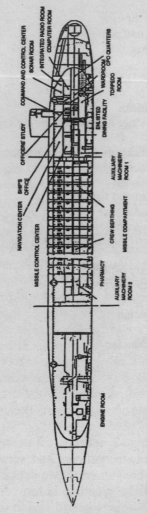

USS NEBRASKA (SSBN 739)
Interior Diagram

AFT

FWD

COMMAND AND CONTROL CENTER
INTEGRATED RADIO ROOM
COMPUTER ROOM
SONAR ROOM

OFFICERS' STUDY

WARDROOM
CPO QUARTERS
TORPEDO ROOM

SHIP'S OFFICE

ENLISTED DINING FACILITY

NAVIGATION CENTER

AUXILIARY MACHINERY ROOM 1

MISSILE CONTROL CENTER

PHARMACY

CREW BERTHING

MISSILE COMPARTMENT

AUXILIARY MACHINERY ROOM 2

ENGINE ROOM

• Prologue

Ray Chesney wandered through the command and control center of the USS *Nebraska* like a car buyer browsing through a showroom. The young Navy ensign ran his hand admiringly over the large, gray steel box in the middle of the room containing the commanding officer's display console, which was used for tracking enemy submarines. Above it hung a roll of toilet paper, utilized to wipe grease-pencil markings off the plastic chart boards.

The command and control center—"control" for short—is the nerve center for the Trident submarine. Control is as large as a roomy bedroom. But with all the instrument panels and stools and chart stands and plotting tables arranged inside it, all the silver tubes and gray pipes and black wires and red phone boxes that hang overhead, not to mention all the crewmen who crowd in to operate its equipment, control becomes cramped.

The *Nebraska*'s command and control center was deserted for the moment; the officers and senior enlisted men of the crew were gathering two decks below for one last early-morning meeting before the *Nebraska* set sail. The submarine was still tied to the pier at its home port, the Kings Bay Naval Submarine Base,

a heavily guarded military facility along a resort section of the south Georgia coast. But within hours it
would begin a long lonely patrol, almost three months
under the Atlantic Ocean. During its first week at sea
the crew would test itself and the *Nebraska*'s complex
machinery to make sure both were ready for war.
Chesney expected it to be the most intense week of his
life.

He stopped in the middle of control, to take in once
more everything he saw. This was the room where the
giant ship was steered, where the captain peered at the
world outside through the periscope, where buttons
could be pushed to fire torpedoes or a key could be
turned to launch, God forbid, the twenty-four long-
range ballistic missiles.

Those twenty-four missiles sat in giant tubes in the
Nebraska's belly. Chesney had thought little about the
damage they could do if they were ever launched. Who
in his right mind could grasp the unimaginable destructive power of the nuclear warheads that sat on top
of those missiles?

The pride and joy of those warheads—at least for
the United States Navy—was the W-88. The Russians
were terrified of it and the Chinese had launched an in-
tensive spy operation to steal the secrets of how it was
made. For good reason. The W-88 packs an explosive
wallop of 475 kilotons, the equivalent of 950 million
pounds of TNT blowing up at once. Hurled to its target
by its powerful missile, the warhead is accurate
enough to plop into a football stadium six thousand
miles away. Depending on the detonation altitude, the
hydrogen bomb in its core could destroy a hundred-
square-mile area, about three times what the atomic
bomb dropped over Hiroshima obliterated.

The center of its blast, "ground zero" as nuclear theologians reverently term it, would reach several tens of millions of degrees within a millionth of a second, vaporizing rocks, soil, and humans that were sucked into a giant fireball. Farther out, the explosion's thermal energy would create horrific firestorms that would char skin, blind eyes, light up homes like matchboxes. A fraction of a second after the explosion, a high-pressure shock wave would blast out and, within ten seconds, flatten buildings three miles away. For another nine miles out, the blast wave would turn bricks, glass, and wood into lethal shrapnel ripping through bodies caught in the open. Those not struck by the shards would suffer massive hemorrhaging, punctured eardrums, or ruptured abdominal and thoracic walls from the environmental pressure of the blast wave.

The nuclear radiation from the explosion would spray deadly neutron and gamma rays that would leave humans vomiting and dizzy in the beginning, then later wishing they had been at ground zero and been spared a slow death from skin hemorrhages, internal bleeding, seizures, prostration, and diarrhea. The longer-term radiation fallout could contaminate skies for a thousand square miles. Inhabitants below could look forward in the years ahead to a buffet of ailments: cataracts, leukemia and other forms of malignant cancers, retarded growth, or genetic time bombs for future children.

All told, just one of the *Nebraska*'s W-88s dropped, say, on Moscow could cause as many as 360,000 deaths and more than one million injuries, most from ugly burns. It was almost impossible for Chesney to picture that kind of carnage—and from just the first shot fired.

The Navy has eighteen Trident ballistic missile submarines like the USS *Nebraska* in the Atlantic and Pacific oceans. Each has twenty-four of the long-range missiles that could propel their independently targeted warheads for thousands of miles to strike anywhere in the world. Atop each rocket is, on average, a mix of five W-88 or W-76 warheads. The W-76 is an older, less powerful nuclear weapon, with only one hundred kilotons of explosive force. So *only* 121,000 Muscovites could expect to die from it, with another 375,000 likely injured. Totaled up, each Trident submarine carries about 120 strategic nuclear warheads, which have twice as much explosive energy as was detonated by all the conventional weapons in World War II. They could destroy every major urban center in Russia, killing or wounding about ten million people.

The Navy started placing ballistic missiles in submarines in 1959. The decision was made more for bureaucratic reasons than strategic ones. The admirals didn't want the Air Force, which already had strategic bombers to attack the Soviet Union and was developing intercontinental ballistic missiles, to hog all the nuclear weapons. As it turned out, the Navy's submarine-launched ballistic missile force became the most valuable leg of the "strategic triad," because, as was not true of bombers in the air or missile silos on the ground, there was practically no way the Soviets could find the subs underwater and attack them before they launched their missiles.

The first Polaris rockets placed in subs were inaccurate blunderbusses, good only for flattening cities. By 1990, however, the Navy had put D-5 missiles in its Trident submarines, and their electronic guidance systems were accurate enough to place the W-88 warhead within 133 yards of its target—a capability that makes the Kremlin nervous to this day, because the American Tridents can quickly knock out Russian missile silos.

Chesney stood lost in his own thoughts, scanning the forward section of the *Nebraska*'s control center. He was unofficially designated the "bull ensign," a title bestowed on the more senior of the two newly minted Navy ensigns who had joined the submarine just two months earlier. Seniority is determined by the date an officer receives his commission. Mark Nowalk, the other ensign on board, was junior to Chesney.

On each of the gold bars pinned to Chesney's collar was engraved BULL. It was hardly a rank of distinction. Nowalk actually had more time in the Navy than Chesney, having served as a sailor for ten years before becoming an officer.

No matter; both were low on the food chain. For a submarine officer's first four years in service, the Navy invests far more in training him than it gets back in productive work. A new ensign on board is only a step above worthless, the reason they are all given another nickname, "nub," for "nonuseful body." Chesney and Norwalk were both assigned easy divisions to supervise with experienced chiefs in each to baby-sit them and with menial duties the more experienced officers dumped on them, like shredding classified documents no longer needed or getting up in the middle of the night to verify the charge lineup for the ship's garage-sized battery (a chore that involves walking all through the sub to check more than fifty gauges and valve positions affected by the recharging).

Chesney was twenty-three years old, but looked five years younger. His wispy blond hair fell back from a wide forehead and a face that always seemed freshly scrubbed. He had grown up as far from the sea and submarines as anyone could get: Brooklyn, Michigan, a tiny farm town southwest of Detroit. He was the second-youngest of eight children; their father was a plumber. Brooklyn had two stoplights and three fast-food restaurants and that was about it. Only 130 students attended Chesney's high school, where he played football and excelled in math and science. It had been a *Happy Days* life. Heather, the girl he would later marry, had been the homecoming queen. Ray had been the homecoming king.

Chesney paid his own way through college, graduating at the top of his class with a chemical engineering degree and $21,000 in student loans. He decided to join the Navy after reading a service brochure about the exotic nuclear reactors that powered its ships. That seemed to him like the best place to apply his training. He was the first in his family to enter the sea service. The thirteen weeks of Navy Officer Candidate School in Pensacola, Florida, had been a shock—Marine Corps sergeants with arms that seemed to Chesney as big as his thighs screaming at him constantly, drills day and night, meals wolfed down while sitting at attention, no sleep, nitpicking inspections.

But nothing in college or OCS or later the Navy's submarine school had prepared him for the raw technological power he now silently inspected. Along with its seventeen sisters, the *Nebraska* was the most complex military machine mankind has ever produced, a $1.8 billion marvel containing more modern technology than any other vessel in the world. Many countries have submarines, but most are for show. Few can operate them at sea for more than a few days at a time. The *Nebraska*, which stayed at sea eight months a year, was the product of a capital and technological infrastructure that took fifty years and $4 trillion to build. The three-stage rockets in its missile tubes each weighed 130,000 pounds and cost about $50 million. The sub could steam underwater for decades, powered by a nuclear reactor with a uranium core no bigger than a subcompact car. (The *Nebraska* actually stayed under no more than eighty to ninety days at a time, because it didn't have enough room for food to feed the crew much past that.)

Mastering all the technology in this behemoth would take years. Chesney was proud that he had at least memorized the position and purpose of every piece of equipment in the control center before they set sail. To the right of the commanding officer's display console box that he leaned on was the digital display indicator console to monitor sonar readings when an enemy vessel was being tracked. To his left stood the captain's indicator panel that the skipper used during the launch of the Trident's missiles, along with the new TAC-4 computer Chesney was just learning to operate, whose multicolored screen could display the movement of the *Nebraska* and enemy vessels.

On another row of hardware in front of him and to the right were the weapons control, torpedo launch, and pri-mate consoles. Operators squatting in front of them on stools calculated the location of enemy vessels, then punched in the directions and armed the fuses for the torpedoes that would be fired at them. To their left was the large ship's control panel, painted black on its front, with row after row of digital and analog indicators and dials, levers and buttons to fill and empty the sub's massive ballast tanks, and colored warning lights for devices that monitored every part of the vessel. Sticking out from the right side of that panel were two large steering wheels that looked like airplane yokes, which young sailors strapped into seats facing them pushed and turned to steer the sub.

To the left of the ship's control panel stood a vertical chart board with a brown imitation-leather cover draped over it. "Ship's Intel" and the *Nebraska*'s blue-and-red crest were printed on it. The cover concealed secret intelligence on foreign vessels in the waters the sub would patrol. To the left of it was another vertical chart, the contact evaluation plot, on whose rolling graph paper a sailor kept a written history of the *Nebraska*'s position in the water and the positions of other vessels near it.

Forward of the control room were the small sonar shack where sonarmen listened to the faint noises of other vessels in the ocean and a larger communications room full of radios and computer closets to receive alert messages from the U.S. Strategic Command.

Chesney pivoted around slowly for a long look behind him at the two periscopes mounted on what the crew called the "conn," a raised half-moon platform with a metal railing around it. Scope one to his left, two on the right. Fat gray tubes, each with rubber eyepieces and foldout crossbars with grip handles to turn the scope around for a 360-degree view of the outside. Crammed behind the periscopes from left to right were a chart table, a depth sounder transmitter and receiver, a shore phone and amplifier, a bearing repeater indicator, an electronic surveillance system console to detect outside radar emissions, and a command and control switchboard. Behind this hardware, separated by a bulkhead, was the ship's navigation center, another room just as large as control, with long tables for preparing charts and computer terminals filled with secrets on where the sub could travel.

Chesney already knew that the conn, where the periscopes were mounted, was a special place. Under way, crewmen had to ask permission to step up on the platform and walk across the conn, to show respect for what was considered Dave Volonino's territory. He was the captain of the ship.

Chesney had a lot to learn about customs and life aboard a submarine. Its world was so different from the civilian world he had left just two years ago. On patrol, for example, submariners live by an eighteen-hour day: six hours on watch, six working in their departments, six off. They claim they adjust quickly to this circadian cycle when they submerge, although psychological studies show they do lose sleep and become moodier.

The *Nebraska*'s leadership was almost lily-white. The lower enlisted ranks had about the same percentage of minorities as in the civilian world, but there were only two African-Americans among the *Nebraska*'s thirty-three officers and chiefs, a demographic found in the upper ranks of other subs. Most of the *Nebraska*'s crew came from broken or troubled homes, also a phenomenon found in other subs. The men with whom they shared this steel womb for months on end would be the family they never had growing up.

And as in a family, there were no secrets on the *Nebraska*, Chesney would quickly discover. No privacy. Enlisted men sized up officers within a day. Gossip spread quickly. Everyone lived in everyone else's dirty laundry, literally as well as figuratively. Everyone knew everyone else's business. Everyone watched everyone else.

For good reason. A screwball nobody has paid attention to is too dangerous to have aboard a vessel that carries nuclear weapons. In addition to their top-secret security clearances, officers and enlisted men who work closely with a Trident's missiles are subject to the PRP, the service's Personnel Reliability Program. Their personal lives and medical and financial records are closely and regularly screened. Any sign of abnormal behavior—a drunk-driving ticket, mood swings, lying to superiors—and a crewman is "taken off the PRP," which means he can no longer be involved in the handling or launching of the missiles. An officer taking cough medicine with codeine can even be temporarily decertified.

Military bearing and chain of command, which can be as rigid as Indian castes aboard the Navy's surface ships, were relaxed on the *Nebraska*. Although criticism could be blunt, screaming was considered bad form. The skipper of a Trident sub in the Pacific was relieved of command in 1997 for verbally abusing his crew. A Trident's officers make requests as often as they give orders to enlisted men. On patrol, *Nebraska* sailors could grow beards, and they wore "poopie suits," comfortable blue jumpsuits whose only indication of rank were the insignias on collars and the belts wrapped around the waist (tan canvas for officers and chiefs, blue for sailors). Crewman also were allowed to wear civilian deck shoes or sneakers instead of clunky military boots. They were more comfortable, and quieter when climbing around the sub's insides. Keeping up morale was important for the *Nebraska*'s high command. Its temperature among the crew was taken constantly. The last place the Navy wants angry men is aboard one of its ballistic missile submarines.

The *Nebraska* was a far cry from the World War II–vintage subs. Chesney didn't feel claustrophobic inside it. Who would? The 18,750-ton steel monster was 563 feet long; stood up, it was taller than the Washington Monument. At its center, a Trident is over four stories high and wider than a three-lane highway. When surfaced, it has a draft almost as deep as an aircraft carrier. Sailors call it an underwater cruise ship. (Russia's Typhoon-class missile sub is even more massive, with so much room inside that it has a pool and aviary for the crew.) The only place that felt cramped in the *Nebraska* was the sleeping compartments for its sailors, each one a twenty-seven-inch-wide, eighty-inch-long, and twenty-one-inch-high box, with a fan blowing in cool air. They were like air-conditioned coffins.

The Cold War ended when Chesney was in junior high school. The Soviet Union shattered into fifteen countries, most of which have a per capita gross domestic product no larger than the Dominican Republic's. Moscow now has a rowdy press, and American political consultants have trekked to the capital to advise candidates on how to manage their political campaigns in freewheeling elections. Russia and the United States are now considered allies, though wary ones; their soldiers serve together as peacekeepers in the Balkans. The Kremlin barely manages to keep one ballistic missile submarine underwater near the Atlantic Ocean and another in the Pacific; the rest remain tied to their piers, short on supplies and collecting barnacles on their hulls. The threat of global nuclear war—remote even during the Cold War because it assured mutual destruction—is now practically nonexistent.

But the Tridents' routine, which Chesney was now learning, changed little with the end of the Cold War.

Every two weeks, like clockwork, an American ballistic missile sub still slips quietly away from its pier and sinks deep into the dark ocean to relieve one of the ten Tridents that remain on constant patrol in the Atlantic and Pacific.

The politicians may be friends, but the generals are not. Russia still conducts mock attacks against the West with its nuclear forces, enfeebled somewhat by lack of funds, but still potent enough to wipe out the United States. Its subs that are tied to piers can still fire their ballistic missiles in port. In fact, because it has no money to pay for the vast army of tanks and warplanes it marshaled during the Cold War, the Russian government now relies on cheaper-to-maintain nuclear weapons even more for its defense—nuclear weapons that remain on hair-trigger alert. The Chinese have a small nuclear arsenal as well, the U.S. generals point out, and hostile countries such as Iran, Iraq, and North Korea are bent on acquiring atomic arms.

On the American side, the ten patrolling Trident submarines now carry the bulk of America's strategic nuclear deterrent, more than a thousand warheads invulnerable to enemy counterattack because the ocean hides them. They are warheads on hair-trigger alert as well, ready to be fired at a moment's notice.

The underwater arsenal is more than enough to satisfy the top-secret targeting plans the Pentagon still has locked in its vaults to destroy eight hundred choice locations in Russia: from Moscow to St. Petersburg, from the Kozelsk rocket forces in the northeast to the Uzhur missile fields in the south, from the Murmansk sub base in the Barents Sea to the Vladivostok naval facility on the Pacific coast. More than enough to wreck already deteriorating Russian industrial production, collapse its decaying medical care, rip apart a social fabric frayed as is, halt a perpetually clogged transportation system, foul vast areas of moderately productive agriculture, and, depending on which climatologist you believe, spread radioactive contamination across much of Europe and Central Asia to unleash a global cooling of the atmosphere from the dark and dusty radiation clouds that would billow up. In other words, it's enough to destroy a big chunk of the earth.

Only the President of the United States can authorize the use of America's nuclear weapons. That order, written in special code, would be transmitted by a low-frequency radio signal to the *Nebraska*, where fifteen officers and 147 enlisted sailors and chiefs would turn the keys, punch the buttons, and finally pull a trigger to fire their missiles. The average age of the *Nebraska* men entrusted with blowing up their part of the world was just twenty-four years. Their captain had been thirty-nine when he took command of the ship.

Although the Cold War has been won and the Tridents' purpose is now more difficult for the Navy to explain to recruits, the service's best and brightest men, like Chesney, still volunteer for this sub duty. (Women, who have broken into the surface ship and aviation ranks, are still barred from submarines.) All of *Nebraska*'s officers had been top students in their class, most with undergraduate and advanced degrees. Most had been techno-nerds in high school who then majored in engineering in college. Almost all the chiefs had at least a bachelor's degree, while most of the sailors under them were working toward one. Submariners over the years have cultivated a rakish image; in World War I, militaries considered their captains just a step above war criminals because they sank ships while hiding under the sea. But most on the *Nebraska* were shy introverts.

The USS *Nebraska* was commissioned in 1993. According to the Navy's official designation, it is an Ohio-class sub, SSBN-739, although neither the hull number nor any other markings are painted on its black skin to identify it to enemy spy satellites. The ship, however, takes its name seriously—too seriously for the tastes of other sub crews in the fleet, who snipe that the *Nebraska* goes overboard with all its state pride. The boat is filled with University of Nebraska memorabilia: glass-encased footballs from championship seasons, framed game photos, helmets, pennants, banners painted red for the school's color. Nebraska politicians are routinely invited to the sub for day cruises, and the Big Red Sub Club of Nebraska often flies the boat's seamen to the state, where they are taken to rodeos, served free drinks in bars, and treated like royalty in parades. Nebraskans, the crew likes to say, are proud of three things: corn, a football team, and their own ballistic missile submarine.

Chesney looked at his watch. No more time for daydreaming. He sprinted down a stairwell on the port side of the control room. He didn't want to be late for the officer and senior enlisted meeting two levels below in the enlisted mess.

For almost three months underwater on lonely patrol, the *Nebraska* would roam an area of the Atlantic as large as four Georgia-sized states. Of those aboard, only the officers and a few enlisted men would know its exact location. It would sail quietly to prevent hostile subs or ships from detecting and destroying it before missiles could be launched. No Soviet attack sub ever found a Trident during the Cold War, its submariners boast. Nowadays no unfriendly vessels even try.

Trident crews, however, still spend long hours training for the unthinkable. Even after submarine school, a Trident's newest members like Chesney put in a year or more of apprenticeship on the boat before they are allowed to wear the coveted "dolphins" insignia— gold-plated for officers, silver-plated for enlisted— which identifies them as qualified submariners. (In the old days, when a seaman was awarded his dolphins, its metal spikes were pounded into his chest—a barbaric ritual Marine Corps paratroopers also practiced. It is now outlawed.)

After serving on strategic alert in a Trident, a seaman also is allowed to attach to his uniform just below his dolphins a silver "Patrol Pin," similar to the one given submariners after they saw combat during World War II. Some submariners believe it amounts to a sort of posthumous award, the assumption being that if a Trident ever fired its missiles out of the ocean in anger, retaliating Russian forces would likely be able to pinpoint the vessel and quickly sink it. Even if the Trident escaped and made it back to Kings Bay, there would be no one alive there after World War III to present the Patrol Pin. So by the logic only the atomic age could bring, the award is given ahead of time.

All of the *Nebraska*'s crewmen accept this Strangelovian pact. It may be hard for outsiders to fathom, but for most of them the belief remains fervent that the holocaust machine they sacrifice so much in their personal lives to sail is still as critical to keeping the world safe today as it was during the Cold War. Fervent for most, but not all. Around holidays some privately grumble and question why the Navy still insists on sending them to sea, away from their loved ones, when the chance of having to launch is almost nil.

They trained and trained and trained, nonetheless, with a sense of mission different from other warriors. A pilot or infantryman perfects his skill, professing not to want war but secretly lusting for the day he might test himself in combat. A Trident submariner trains for a war he genuinely hopes he will never have to wage. Fire your missiles and you might as well steam to Australia, the submariners like to say, because your family and friends will be turned into glowing ashes after the Russians retaliate against Kings Bay. Nevertheless, each time the *Nebraska* sets sail—as it was about to now—the crew takes grim pride in the fact that its vessel is the sixth-largest nuclear power in the world.

Go, Big Red.

PART I

• • •

THE FIRST WEEK

1 · Going to Sea

About fifty officers, chiefs, and senior petty officers filled the seats in the three rows of dining tables in the crew's mess of the USS *Nebraska*. They were the third of the crew who supervised the rest of the men. The crew's mess, on the sub's third level, was one of the more spacious rooms in the submarine. The plastic, red-checkered tablecloths had now been taken off the thirteen fake-wood-topped tables bolted to the deck. Along the starboard and aft bulkheads were mounted a large-screen television set and VCR, plaques, glass boxes with University of Nebraska footballs, and posters, plus lockers with the some six hundred movie videotapes kept on board for the crew to watch. "Cornhusker Cafe" was engraved on a wooden sign hanging over the galley forward of the dining area, where the food was prepared and served. A salad bar stretched down the middle of the galley's serving area, and to the right was a soft-drink dispenser and coffeemaker.

David Weller now ran this meeting. He was a master chief petty officer whose title was chief of the boat, or COB. He was almost always called "Cob," rather than by his last name. Weller hardly fit the caricature of a beer-bellied chief with an anchor tattooed on his arm. Tall and lanky, with a blond mustache and thick glasses, Weller spoke softly and was an astute manager. He wielded enormous power on the submarine. He was part of the *Nebraska*'s senior management. Officers thought twice about challenging him. He was nearly equal in status to the sub's executive officer. Weller was the man who commanded for the most part the enlisted ranks and represented their interests before the captain.

Just twenty-four hours earlier, the COB had been in a foul mood. Senior officers and chiefs from the *Nebraska*'s parent squadron were coming aboard to inspect the vessel during the first week of its patrol, and the sub looked like a teenager's room. Instead of working together to complete the final repairs and get under way, the departments and divisions "were acting like eighteen fucking unions," he had growled. To top it off, two brand-new red T-shirts had been snitched from one of the ship's lockers.

Thursday morning, however, everything had finally come together. It always did, Weller knew. Tools and gear had finally been stowed, and for hours the day before, the entire crew had stooped to its hands and knees to scrub passageways and ladders. Instead of mops, they had used sponges and towels on the floor, because mops retain water and in the sub's enclosed environment the stale smell is hard to bear.

Now the COB had one last item on his checklist: the muster. Submariners live by many important rules—among them, surface as many times as you dive and don't leave port with any of your men standing on the pier. A submarine like the Trident has no fat in its roster, not like an aircraft carrier, which is a small city of five thousand seamen. Because the *Nebraska* was so complex and the crew numbered only 162, each man had multiple duties and chores. A storekeeper on one shift could serve as a helmsman on another. A reactor operator might also serve as a diving officer. Everybody did windows, and absences were painfully felt. Weller wanted to make damn sure his chiefs and petty officers had accurately counted all their men present before the *Nebraska* cast off its lines.

The supervisors assured him everyone was accounted for. "All right, let's go kick 'em in the ass," the COB finally said.

The submariners broke out of their seats, laughing and joking, ready to do just that. The crew was in a good mood this morning. Spirits always seemed to brighten the day they were supposed to get under way. But Chad Thorson remained in his chair for a few moments longer. He had sat quiet and sullen through the entire meeting. Thorson couldn't think of anything to be cheery about. He was leaving his beloved Kyung, disappearing from her life for almost three months.

"Stick" felt miserable. That was the nickname the other officers had pinned on Lieutenant Junior Grade Chad Thorson. He had delicate features and a shy, soft voice. He wore glasses and was brainy and skinny as a rail—five feet eleven inches and 140 pounds. His weight was all the more remarkable for the fact that he ate like a horse when he was on patrol. He never gained so much as an ounce. Thorson filled his tray to capacity at each meal, breakfast, lunch, dinner, and the midnight rations called "midrats," and wolfed it all down. As an officer he was charged $500 for the meals he ate on patrol, so he intended to get his money's worth. He considered the eighty-six-day cruise an all-you-can-eat buffet.

When he came aboard the *Nebraska* a year and a

half ago as a timid ensign, Thorson and the bull ensign who reported with him had first been nicknamed "thing one and thing two." But the *Nebraska*'s other officers were convinced that a tiger stirred inside the heart of twenty-five-year-old Chad Leif Thorson, just waiting to get out. Thing two, Thorson's unofficial rank, could be a vaudevillian during crew skits. He could scare his shipmates to death by standing up to the biggest bully in a bar.

His grandparents, who along with his divorced mother had reared him, had pressured Chad to follow the family tradition and become a Lutheran minister. But he could never see himself as a man of the cloth, although through most of high school he had played the good grandson and promised to follow the family tradition. When he turned seventeen, Chad had persuaded his mother to sign the permission papers that allowed him to join the Navy. He was desperate as well to escape the farming village of Courtland in south-central Minnesota; if he didn't do it then, he knew, he'd be stuck there forever as his three sisters still were.

Navy boot camp in hot, muggy Orlando, Florida, was a culture shock for a fair-haired boy who until then had never set foot out of chilly Minnesota. But Thorson was thrilled with cutting the umbilical cord and took well to the discipline of military life. He soon was invited to train to become an officer at the U.S. Naval Academy in Annapolis, Maryland. The thought of it terrified him at first. A pipsqueak among these big midshipmen with their deep voices, he'd never survive.

Thorson soon discovered Annapolis wouldn't kill him. He even excelled academically, making the superintendent's list with a 3.6 grade average in electrical engineering, the killer major at the academy.

And along the way, he fell in love with Kyung Lee.

It was a cyber-generation courtship. They first met over the Internet and for a month exchanged e-mail every day in an electronic chat room. Kyung, who was born in South Korea and had moved to Wilmington, North Carolina, when she was eight years old, was a pharmacy student at the University of North Carolina at Chapel Hill.

Thorson finally drove down to Wilmington to meet the girl whose picture he'd never seen but whose e-mails had captivated him for more than a month. Kyung had long black hair and deep brown eyes and was at first shy, like Thorson. He fell in love with her almost instantly.

From then on, every weekend he could get a pass to escape the academy, he made the long drive to North Carolina to be with her. Their romance had been different from others. The talking had come first, then there was the physical attraction. Before they even saw each other, they had communicated volumes by e-mail, and after they met they continued to be ardent electronic letter writers during the long separations while both finished college and Chad went through submarine training.

That was why Thorson was aching now. Within hours, the *Nebraska* would set sail, and for almost three months he would not be able to communicate with Kyung. Few personal messages from shore got to the crew when the sub was deep underwater, and none could be radioed out because it might let the enemy know the vessel's position. Letters had been what began their romance and kept it alive when they had been separated in the past, but patrols at sea meant they were practically divorced.

Thorson had picked submarines when he graduated because it was the most exclusive club to join. He had not finished challenging himself. But the first time he had to leave Kyung to sail on an eighty-day patrol—they were engaged by then—he cried the night before the sub pulled out. He couldn't talk to her every day. He didn't even know if she had made it back to Chapel Hill safely after she dropped him off at the pier. That had been agony for him, and when he returned from his first patrol he discovered that Kyung had hated him for putting her through that kind of loneliness.

The second eighty-day patrol had been more bearable. The two learned to cope better with the separation. Now, for the third patrol, Thorson had a shiny new wedding ring on his finger. He had married Kyung just two and a half weeks before. They had returned from their honeymoon ten days ago. And now he was abandoning her again for another eighty-six days.

Chad had remained on the boat all of yesterday, because he had been stuck with the job of serving as one of the duty officers. He and Kyung had said their good-byes earlier. But late last night, a security cordon around the wharf had been lifted and wives and girlfriends had been allowed to board the sub for a couple of hours.

Chad was elated, Kyung not so much. They had developed different personalities. Kyung was far more independent and assertive now than when she first met Chad. When it was time for him to leave, Chad became sentimental and clinging. But Kyung wanted to start separating earlier. If he was going to be out of her life for eighty-six days, she wanted to make the break quickly and get on with living alone. She had climbed down the long metal ladders last night to be with Chad inside the sub for two hours, but she didn't like having to go through a second goodbye.

Some of the officers and chiefs remained in the crew's mess, opening up schedule folders to review the events for that morning. Others had rushed to the stations they would man when the sub got under way. Thorson finally stood up from the mess table and ambled to the back of the sub. His job for this patrol was being the damage control assistant. He watched over the division of men who maintained machinery and hydraulics throughout the boat. On previous patrols he had worked with the reactor. He liked being in the reactor section in the back of the sub. The captain didn't visit there often, so he wasn't in your face much. Volonino let the sub's engineer pretty much run the show back there.

Thorson had made a decision he had not yet shared with the captain or any other officer on the sub. He was getting out.

The *Nebraska* was a big family, and, except for a few whiners in the engineering department, Thorson thought they were the best bunch of guys he would probably ever work with. Volonino certainly was the best skipper he could hope for—very competent. Everybody knew that and felt good about it. He almost never yelled at people. In fact, it would shock Thorson if he did. Volonino was a product of the Navy's new leadership, which believed you got just as much out of a sailor by explaining to him and motivating him. A few of the old screamers who ruled their ships by intimidation were still around. He had talked to officers on other subs at Kings Bay and heard the horror stories. *Nebraska* had the reputation for being the best boat on the waterfront.

But Thorson had no intention of putting Kyung through these separations for the rest of their working lives. She had said she could handle his going to sea, but he knew she hated it, and so did he. Being married meant being married, not being away from your wife five months of the year. It was bunk, the lines Volonino and the other veteran officers fed him about how great it was to be out at sea and how the wives eventually adjusted. He'd seen all the divorces in the submarine service. The couples who stayed together and claimed they didn't mind the long separations probably shouldn't have been married in the first place. Certainly they shouldn't have kids. When his service contract with the Navy was up in May 2001, Thorson was leaving.

Minutes before 7 A.M., the command and control center came alive. Like stage workers appearing from nowhere to light up a set and begin a play, sailors and chiefs and officers poured in and began hooking around their necks phones (whose communications are powered by the sound waves of their voices) or fitting pen mikes and earpieces on their heads. They wiped off the Plexiglas on control panels for the last time, flipped open operator manuals, turned switches. All the banter and chatter from earlier in the morning disappeared, now replaced by quiet efficient movements. There was little talk. Everyone knew what he was supposed to do. No need to discuss it. For the next two hours, the men in control would be guiding the submarine through a narrow winding channel to a point in the Atlantic Ocean deep enough for the vessel to submerge. It was one of the trickiest sailing maneuvers the sub had to make on its patrol. Petty officers from the

ship's navigation department spread out charts of the
Kings Bay harbor and the channels that snaked out
from it, and taped them down on primary and secondary plotting tables. Pencils and calculators and rulers
and dividers and protractors were set out.

Tom Standley, the senior chief of the navigation department, hunched over the primary plotting table. He
had bulging eyes, buck-teeth, a bushy mustache, and a
southern drawl from his native Danville, Virginia. He
was one of the best navigation chiefs Weller had ever
seen, meticulous with the charts. He also kept a firm
grip on the sailors in his department, who sometimes
could be a rambunctious lot. But this morning, Standley had a throbbing jaw. Just his luck, the day before
the ship was to leave, he had to visit the base dentist.
"Goddamn Navy dentists," he now grumbled. The
sonovabitch had been about as delicate as a hard-hat
with a jackhammer, cutting all that gum just to pull one
damn tooth. He had wanted to yank out another, but
Standley had said no way. He'd still be paralyzed with
Novocain in that dental chair when the *Nebraska*
shoved off.

Lieutenant Al Brady, the sub's navigation officer,
paced behind Standley with a headset receiver and pen
mike that almost touched his lips, its cord attached to a
walkie-talkie clipped to his belt. "Al Brady the lion-
hearted," as Volonino liked to call him. The hardest
worker the skipper had ever commanded. What Al
lacked in raw talent he made up for in personal drive.
He was a perfectionist, very focused, just like his wife,
Jenny. Wives tended to reflect their husbands' personalities, Volonino had found. As the navigation officer,
Brady was a department head, a step above the sub's
junior officers. He wanted to make the Navy a career,

perhaps one day command his own sub. Jenny, a teacher, wanted that for him, too. She could rattle off Navy lingo as easily as her husband did.

The bustling continued. Lieutenant Commander Alan Boyd, the *Nebraska*'s executive officer, patrolled control with a walkie-talkie in his hand and a scowl on his face, like a wolf ready to pounce on any mistakes he saw in the preparations. Boyd would be in charge of the room while Volonino guided the boat from above on the bridge. The XO would spend much of his time during the trip out of the channel watching the weapons control console on his left, which monitored the ship's heading, and the digital display indicator console to his right, which flashed the radarscope. Boyd was new to the sub but already questioning the way its junior officers did things and brusquely second-guessing them. He was determined to run the control center his way whenever he was in charge.

Brett Segura, a petty officer first class who usually was up to his elbows in the sub's machinery, sat down on the chief of the watch chair, to the left of the young sailor who would drive the boat. Segura began twisting knobs and punching buttons on the complicated panels that controlled the sub's ballast and trim and hydraulics and missile hatches and communications circuits.

Steve Habermas, a lieutenant who was the sub's communications officer, plopped down in the diving officer's stool to the right of Segura, just behind the seats for the two sub drivers, to confirm that the ship control panel settings were in the proper position to get under way. A Naval Academy grad, Habermas had done his master's research project on digital signal processing. When they pushed away from the pier, he would be the control room supervisor, and he would handle administrative chores during the sail out the channel.

Matt Suzor, a petty officer second class whose wife would deliver their third child while he was away on this patrol, scribbled with an electronic pencil on an IBM ThinkPad the size of a book. It was the ship's log. The record of all the ship's movements was now kept on a computer instead of in a notebook.

Lieutenant Ryan Hardee, a nuclear weapons officer, walked through carrying a cotton American flag that had been flown over the old USS *Constitution*. The flag's owner had asked that it be flown from the *Nebraska* during its patrol, and Hardee had promised he would hoist it. Hardee, who was born in Dimmitt, Texas, near Amarillo, was a prankster among the officers, but a talented tactician during battle stations. He was cocky and handsome. The other junior officers ribbed him about being vain—always checking his hair in a mirror and clipping his nails.

"Sonar, radio, nav center," Todd Snyder, the sonar section's new chief, commanded. "Test the open mike." Sailors pushed buttons on handheld microphones strung from the ceiling on coiled black phone cords and called out communication checks. All the internal ship's alarms were sounded to make sure they worked: the collision alarm, general emergency alarm, missile emergency alarm, power plant casualty alarm, diving alarm, missile jettison alarm. They whistled, honked, clanged, and buzzed, all as they should.

Dave Volonino walked in briskly.

"Captain in control," the inboard helmsman in the room announced loudly, as he always did when the skipper entered.

Volonino hopped up onto the conn, flipped down the

folding seat between the periscopes that was reserved for him, dropped into it, and began writing notes on a pad. He wore his freshly pressed khaki uniform. On his head was a blue *Nebraska* ball cap with gold "scrambled eggs" on its brim. The officers and chiefs were still wearing their khakis and the sailors were still in work dungarees. Everyone would switch to the blue poopie suits when the sub was under way.

Volonino had brown hair, trimmed closely on the sides and brushed back on the top. He was solidly built and wore silver-rimmed glasses most of the time. He had a razor-sharp mind and a quick wit. His officers thought he was one of the smartest men they had ever served under, and in fact there was no one on the *Nebraska* who knew submarines better than Volonino. He was called "captain" by the crew, but actually he was a Navy commander. At one time the Navy did have captains in charge of its Tridents, but to give younger officers more of the precious command time, they now have commanders, just one rank below captain, skippering the boats. Volonino was now forty-one. But this morning he felt ten years older.

In the last few days, he had been wound tighter than a steel coil. The most irritating parts of his job were always in port, not at sea. He'd grown peevish and snappy, which he rarely was at sea. "If the Navy gave me five thousand-dollar bills and five rounds of ammunition to motivate people, I'd get anything I want," he told his officers at one point, only half joking. "I'd probably use the five thousand dollars every year. The five bullets I'd never use. They'd just be there for deterrent value."

Last night had been the most exhausting, "kind of like the birth canal of life," he said with a laugh. He had slept only a couple hours. Over the years, Volonino had found he could operate on two hours of sleep per night for four days in a row on patrol, but no longer. Then he had to order himself to bed for a full eight hours and let the executive officer handle the problems. He monitored his own fatigue level and that of his men carefully.

Today he was operating on adrenaline, and he'd make a show of being pumped up about going to sea. But he knew his men were dog-tired. The refit had worn them out and left them surly. Volonino didn't blame them. He hated it just as much, what they had been through the past twenty-one days.

To keep them at sea longer, Trident subs were actually manned by two separate crews, designated Gold and Blue. William Porter, a handsome young Navy commander who'd turned prematurely gray and was a close friend of Volonino's, was in charge of *Nebraska*'s Gold Crew. Volonino commanded the Blue Crew. When one crew was at sea with the boat, the other took up duty at the Kings Bay Naval Base, resting, catching up on vacation time, and training new members. The Trident Training Facility on base was one of the largest buildings in the military, with 168 classrooms and labs that could simulate practically an entire submarine. The building even housed mock ballistic missiles in giant tubes, on which sailors could practice maintenance. When Porter's crew docked more than a month ago, it had spent three days turning over the sub to Volonino's Blue Crew. Then the *Nebraska*, as it did every patrol cycle, went through three weeks of what was called "refit," which this time amounted to a mini industrial overhaul of its equipment.

It had been miserable. Sailors were glad to begin

their lonely patrol if for no other reason than to get out from under the crushing load of refit. This time it seemed that all of the sub's insides had been ripped out and repaired. The *Nebraska*'s data processing equipment room—DPER, nicknamed "deeper"—had been completely gutted and upgraded to modernize its computers. Foundation and cable work went on everywhere. The giant diesel generator on board, which provided backup power if the reactor failed, blew a transformer, then shut down from a fresh water leak. The front end of the boat was torn up with noisy, smoky, dirty repairs. The propeller in the back end had to be replaced, because it had developed cracks. On top of that, there was chipping and painting of practically every surface twenty-four hours a day, along with the normal testing, maintenance, and training that had to be accomplished to make equipment and men proficient. Even when he took a day off during the weekend, Volonino's cell phone rang constantly with repair updates.

About six hundred major jobs had to be completed, and Volonino closely monitored each one. The *Nebraska* was such a complex machine that repairs in one part of the ship inevitably threw off operations in others. Welding in one section could create a spark that might ignite fuel oil being transferred in another compartment. Flipping on a circuit breaker to check a coffeepot in one room might shock a worker testing bare lines in another room. During refit, a thousand red tags hung from switches and levers warning sailors not to touch or turn them because it would endanger people in other parts of the boat.

Last week had been a nightmare. Volonino halfway expected it. The sub always ran better at sea than tied up at port. Idle along the dock, it developed all kinds of problems, like a car that's been parked in a driveway too long. Refit could actually be as dangerous as combat for the sub, because all its mechanical and electronic systems were out of whack and there were fewer crewmen on board during off hours to deal with emergencies.

Just before it prepared to set sail, the *Nebraska* seemed to have become infected with what Volonino called "material pop-ups"—bureaucratese for gremlins, many of them in the ship's electronics, which had been powered on and off so many times during refit. A chief who had the most high-profile equipment problem at any time "had the football," and kept it until the next chief faced the next big glitch.

In the last couple days, the chiefs had been handing off the football like quarterbacks. A torpedo went bad (its guidance electronics refused to receive instructions from the ship's computer) and the 3,400-pound weapon had to be gently hoisted out of the sub and replaced with another, a dangerous job that took hours. Then the ship's Global Positioning System equipment went screwy when electricians gave it a Y2K check. And a locking mechanism on the radar mast broke. When the reactor was started up, a steam plant valve operated improperly.

Then came the killer pop-up. As the crew worked its way through a simulated missile firing test and tilted open the seven-ton metal hatches that covered each tube, a tiny L-shaped bar from the missile lock actuating ring on the hatch for tube number twenty-four snapped at its weld. The hatch now wouldn't open automatically. If the sub was at sea and the crew had to launch the missile, the hatch would have to be opened manually, a time-consuming operation. The base repairmen didn't have a replacement for the obscure part. A new one would have to be manufactured, and unless maintenance crews worked day and night, the part wouldn't be ready and reinstalled into tube twenty-four by the *Nebraska*'s departure time.

The *Nebraska* could delay leaving Kings Bay until the hatch ring was repaired, or it could set sail with tube twenty-four out of commission. It would hardly threaten U.S. national security if the sub was a day late beginning its patrol, or if it had twenty-three ballistic missiles that could be fired quickly instead of twenty-four. But the Navy still operates its sea schedule for Trident subs practically as if it were in the middle of the Cuban missile crisis. The brass worries that if too many Tridents are tardy in their departure times, the service might become lackadaisical with strategic patrols. The Kings Bay base scrambled like paramedics on an injured patient to repair the ring. *Nebraska* was the priority, the next sub to get under way for patrol. Within hours of the mishap, the Kings Bay admiral in charge of the ten Tridents that patrolled the Atlantic had a formal briefing on it, complete with a color slide show on the broken part. A team of repairmen from the wharf was thrown at the problem.

Volonino was also obsessed with fixing tube twenty-four quickly. "This mighty warship will get under way on time!" he had told his officers in a booming voice the day before. He had never been late on a departure time. For some patrols, he had even left a day early. Skippers who sailed late got bad reputations on base. Their crews became the butt of local jokes. Being late meant that the sub the *Nebraska* was supposed to replace had to stay at sea waiting for its relief to arrive.

This didn't sit too well with the crew stuck out there. The U.S. Strategic Command, headquartered at Offutt Air Force Base in Omaha, Nebraska, divided up the times each sub was at different stages of alert in the Atlantic or Pacific. It assigned each sub on patrol a "target package," the places in the former Soviet Union or other enemy countries that would be struck with nuclear warheads. STRATCOM demanded that the subs cover all the target packages twenty-four hours a day, seven days a week. No excuses.

Three days ago, Fred Freeland had walked into Volonino's stateroom with top-secret documents spelling out how the *Nebraska*'s missiles would be fired to cover the target package for this patrol. A seasoned lieutenant, Freeland headed up one of the largest departments on the sub. He was almost never called Freeland. Instead it was "weps," the nickname that came with his position as weapons officer. The Patrol Order contained a long list of coded numbers and letters that would be entered into the fire control system for the missiles. The list didn't name places that would be destroyed, but Volonino had long ago figured out the countries that the alphanumeric codes stood for.

It wasn't until late last night, however, that he had been sure that his vow to sail on time could be kept. Tube twenty-four had finally been fixed.

Volonino now finished scribbling notes on his pad and reached up to grab an overhead microphone for the speaker system that could be heard throughout the sub. It was called the 1-MC.

"I want to congratulate the entire crew for a real superb job," he read from his pad, reciting the litany of obstacles they'd had to overcome so the *Nebraska* could cast off its lines at the appointed hour of eight o'clock this morning.

But he had a final warning. "I want to remind you all that when we get under way, we'll have riders aboard," he said over the speaker. The riders were the ten senior officers and chiefs from Squadron 16, who were coming aboard to snoop around the sub during its first five days of refresher training. During refresher training, or REFTRA, as the crew always called it, practically every system on the sub would be tested and the submariners would run through battle scenarios they might face. This would let Volonino and the squadron pooh-bahs determine if they were ready for the patrol. Squadron 16 had the *Nebraska* and four other Tridents under its command.

"Remember the importance of first impressions," Volonino warned his crew. "I don't want to spend the entire REFTRA recovering from a bad first impression."

Particularly not with the leader of this inspection team, Captain Jerry Hunnicutt, who was the squadron's deputy commander. Volonino didn't say that over the loudspeaker. He hadn't heard much about Hunnicutt, who was fairly new to the squadron, but the scuttlebutt was that he was no-nonsense, not a back-slapper. Hunnicutt was also one of a few current officers in the Navy who had commanded both a ballistic missile submarine and a fast-attack hunter sub that fired only torpedoes. Don't try to bullshit him, Volonino had warned his officers. He'll see right through you.

Things sometimes got tricky with a senior officer on board. Even with Hunnicutt there, Volonino was still the captain of his ship. Navy regulations wouldn't allow Hunnicutt to take command, and Volonino guessed he was professional enough not to even try. Hunnicutt couldn't give him orders. But he would be watching over Volonino's shoulder. And he could make suggestions. And Volonino was smart enough to know that he had better follow those suggestions.

Volonino clipped the mike back to the overhead speaker, then bounded down from the conn and headed for the ladder that had been folded down into the center of the control room. He climbed quickly up its slippery metal rungs, through the dank-smelling insides of the sail, in order to reach the bridge at its top, watching all the while not to bang his head or shoulders on sharp ridges that hatches closed down on when the sub submerged. The sail was the giant black hump protruding up near the front of the sub, looking like a big vertical fin—which in fact it was, to stabilize the sub when it traveled underwater. When the sub surfaced, the captain and several crewmen could climb up the ladder inside the sail and step onto the bridge at its top to guide the boat's voyage.

Volonino climbed up through a square door that had been flipped up in the grated metal floor of the bridge. It was then shut down so he could walk over it. The black steel sail now encased him on the bridge, high above the sub's topside.

The bridge had already been rigged for the voyage on the surface. A Plexiglas windshield had been fixed to its front to break the wind. Mounted on the top left ledge of the bridge was a small box for the radar repeater, whose green screen plotted the position of other vessels around the sub. To its right sat the differential GPS box with another screen that gave the longitude and latitude of way points and the sub's position through the channel from signals beamed down by

GPS satellites. A larger box, mounted below the radar
and GPS, was a new gadget electricians had just
brought on board. This commercial SVS-1200 computer had a screen that displayed the sub's position on
a colored map as it sailed out of the channel. Then just
below the SVS-1200 was the Bridge Suitcase, which
housed the collision alarm, a compass repeater, dials
showing the ordered engine speed and rudder movements, and communications circuits with the control
room and engine room.

It was cramped on the bridge. Volonino fitted himself between Freeland and Ed Martin. Freeland was
now the officer of the deck on the bridge. He scrunched
to the right side as he tested a microphone that connected him to the loudspeaker in control. A white canvas bag stuffed with navigation charts sat at his feet.
Martin, a petty officer first class, draped the strap of the
second phone, a sound-powered one, around his neck.
He pasted a plastic sheet to the windshield. Written on
it were the GPS way points the sub was supposed to
cross on its way out of the channel. One of the reactor
operators, Martin had a second job during the sail into
and out of port—bridge phone talker and GPS box operator. His third job was being one of the ship's barbers. And he had a fourth job when not in uniform. He
was a licensed Southern Baptist minister.

Behind the three men, perched a foot up out of holes
in the bridge on the lookout stations, were Ray Chesney and Eric Chambers, both with binoculars strung
around their necks. Chesney was too new to know exactly what he was supposed to do with his binoculars.
Chambers would keep the real watch. A petty officer
third class who was two years older than Chesney,
Chambers had been a skateboarding rebel in high

school. Now he worked in the sonar shack, which had a collection of the closest thing to rebels on the sub. Chambers would call out the buoys the sub passed in the channel and warn if any pleasure boats got too near. Whenever Volonino or Freeland ordered the rudder to be moved right or left, Chambers would also glance back at the large black fin that stuck out of the water and verify that it had turned the correct way.

No one said much now as they busied themselves with preparations to get under way. Freeland began reading off a checklist from a white binder. Martin ran through communications checks on his phone. Volonino climbed up over the lookout station and stood on top of the sail. Poking up from the top of the sail, in the front, was the radarscope that tracked other vessels when the sub traveled on the surface. Behind it were the two periscope tubes. Behind the periscopes were two thick tubes, the sub's multifunction masts that transmitted and received radio messages and signals from GPS satellites. And behind them was the snorkel mast with its mushroom-like top; it could suck fresh air into the sub if needed. Volonino keyed the walkie-talkie in his hand to see if he could talk to the three tugs that were sailing toward the dock, then took a long look around from this highest point on the huge steel beast.

Only a third of the sub was above water when it was surfaced. The other two-thirds remained submerged. Navy divers earlier had swum all around and under its massive hull. They had carefully inspected everything: the *Nebraska*'s smooth sonar dome at its bow, which was made of glass-reinforced polyethylene; the hump on top of it, nicknamed the "shark's tooth," that housed a special high-frequency sonar; the shutter doors for the four torpedo tubes; the four small holes on each side of the hull for launching countermeasure devices; the grates underneath for the six main ballast tanks; the holes in the back where seawater was sucked in to be run through the main turbine condenser; the huge pipes at the end of the rear stabilizers that could string out towed sonar arrays; and the fanged propeller made of a precision-machined aluminum, nickel, copper, and bronze alloy. The divers were making sure that saboteurs hadn't attached bombs.

It was a beautiful clear morning. The rain clouds from the previous night had been swept away. A warm spring breeze now drifted across the sail, and a bright yellow sun rose up off of Crab Island to the east, a manmade island built from the dirt the Navy had piled up after dredging the channel. The *Nebraska* had been tied up to refit wharf number two in the north turning basin, where brick warehouses and yellow loading cranes sat on its dock. The sub was pointed northwest toward the mouth of Kings Bay. Ahead, off its bow, was the giant dry-dock warehouse where Tridents were perched on cement blocks for overhaul.

Behind the *Nebraska* off its stern, another Trident sub was tied up, and farther south stood explosive handling wharves number one and number two, giant gray hangars where the Trident's nuclear-tipped missiles were carefully loaded into the ship's vertical tubes. Six tall lightning towers stood around each hangar.

The hangars were cut short at the end, so the stern of each sub stuck out of the building about thirty feet. That way, Russian spy satellites overhead could see when a Trident went in for loading. Moscow can verify from satellite photos that the number of boats the U.S. Navy equips with ballistic missiles does not violate arms control treaties the two countries have signed. The treaties also allow Russian inspectors to visit the naval base from time to time and peek inside the nose cone of one of the Trident missiles to count its warheads.

The Kings Bay Naval Base is tucked along the Georgia coast just above the Florida state line between Cumberland Island to the north and Amelia Island to the south. The barrier islands have been one of President Bill Clinton's vacation hangouts, and the quaint village of St. Marys nearby is a favorite of tourists.

Tall, barbed-wire-topped fences with warning signs enclose the sixteen-thousand-acre base. Security is always tight but was even tighter now. The base was under Threat Condition Alpha because it was feared that terrorists might try to retaliate for the U.S. cruise missile attack the year before against Osama bin Laden's hideout in Afghanistan. Kings Bay was a choice target for revenge, the only facility on America's East Coast where nuclear weapons were stored.

It has always been peaceful outside the base's fences, however. The south Georgia rural community around it is deeply patriotic, and residents pay little attention to the fact that all those nuclear warheads are in their backyard. Twice a year, on the anniversaries of the Hiroshima and Nagasaki explosions, only a handful of demonstrators gather outside the front gate for a silent protest.

Kings Bay is actually two bases: upper base in the northwest, where barracks, family housing, headquarters buildings, and grocery stores are located, looked something like a high-tech office park with duck ponds, walking trails, a McDonald's, and even a wildlife refuge; lower base to the southeast, with more barbed-wire-topped fences with motion detectors and more security clearances needed to pass through its gates. That is where the subs are docked, where nuclear warheads are stored in underground bunkers and fitted onto the missiles. Four hundred heavily armed young Marines guard lower base with orders to shoot to kill first and ask questions later if they catch anyone trying to sneak into the area. Armed convoys truck the missiles on lower base's nearly deserted roads to the explosive handling wharf for loading. Tall cement guard towers stand along the roads on which the missile convoys drive, with Marines manning machine guns atop them. Surrounding the underground weapons bunkers and the wharves is a hundred-foot-high berm. The hope—and it would probably prove futile—is that if there is an accident in the loading and a detonation occurs, the berm will direct the blast up and limit the ground damage.

One of the three tugs crept up to the bow of the *Nebraska*. A tug worker heaved a heavy line to sailors on the sub's topside, who hitched it to a large cleat painted white. Volonino spotted a crewman who didn't have his canvas safety harness hooked to the topside runners. The harness was to keep him from falling into the water if he slipped off the bow.

"Hook in your safety track!" the skipper yelled.

The sailor did as he was told.

Volonino lifted up the binoculars hanging from his neck and scanned the horizon.

"Warm up the main engines," he finally ordered.

"Warm up the main engines, aye aye, sir," Freeland repeated and relayed the order on his mike. Every order Volonino gave on the sub, even the most routine ones, was repeated by the person receiving it. It cut down on misunderstandings about what the subordinate was supposed to do.

The sub's reactor technicians had begun lifting the

control rods out of the nuclear core that morning to increase its fission rate and generate heat. A Trident is powered by a compact, pressurized water reactor in the rear of the boat, which is encased in thick shields to protect the crew from its radiation. The nuclear propulsion plant has two systems. In the primary system, the reactor heats water pumped through piping loops under high pressure so it won't boil. That hot water then is fed through a steam generator where it causes a second batch of water in the secondary system to turn to steam, which drives the turbine engines.

A second tug nudged up to the right side of the sub's stern.

"Safety quiz, Mr. Freeland," Volonino shouted down to his weps. "What is the proper signal for a ship leaving its berth?"

"One prolonged blast," Freeland answered on cue. They'd had this dialogue before.

"And it will be a long, testosterone-pumping blast," Volonino said with a swagger.

"Yes, sir," he answered, smiling only slightly. The crew had another nickname for Freeland—"the ice man"—because he almost never let his emotions show.

Over the 1-MC speaker, the announcement came that *Nebraska*'s shore phones would be disconnected in two minutes.

A grizzled old harbor pilot from one of the tugs, winded by the climb up, hoisted himself into the bridge, which now became even more crowded.

Volonino trained his binoculars aft to the sub's propeller, which was slowly beginning to move.

"Have I mentioned lately how much I love this job!" he screeched in a voice that sounded like the comedian Jim Carrey's.

It startled Chesney. Freeland, looking ahead, just rolled his eyes and shook his head slightly.

The skipper was always doing things like that to pump up the men before a patrol. And if the truth be told, Volonino probably was the most excited one on the boat about going to sea. No doubt it was because of his divorce, the crew speculated. It had hit him hard, and the sea was his escape for the moment. The men knew that he had gone out on a date the other night, the first one since the divorce, they believed. Maybe that would help. Little of the skipper's personal life was a secret. Gossip about him spread quickly to every man in the sub.

Volonino was both thrilled and nostalgic about setting sail this time, and not just because of the divorce. Even with his hair being on fire the past three weeks of refit, this was the best job he had ever had in the Navy—and the best job he would ever have in his life. He considered commanding a sub to be the pinnacle of his career. He had already been selected for the rank of captain, and his crew believed he would someday become an admiral. But that didn't matter to him now. There had been no greater challenge than leading the *Nebraska*, and it would soon be over. His two-and-a-half-year sea tour was nearly up. A month after this patrol ended, he was to be transferred to a staff job in Washington that he knew wouldn't be nearly as much fun.

God, he would miss these men. More than he could tell them. Never before had he felt so strongly that he belonged in a place as he did with the *Nebraska*. And belonging had become very important to him.

David Michael Volonino was born on July 26, 1957, and reared for most of his life in Apalachin, a small town in south-central New York state near Binghamton. The town was also near the IBM plant in Endicott, where his father, Dominick, worked for twenty-five years as a machine repairman, instructor, then consultant. A first-generation Italian-American, Dominick Volonino was now sixty-nine years old, and still he refused to stop working. He didn't need the money and Dave had tried to talk him into retiring. His father wouldn't hear of it, if for no other reason than that the work kept piling up on his desk. Dominick had taken some college courses but never received a degree. What he lacked in educational credentials he made up for in drive.

The elder Volonino had probably had the biggest impact on his son's life. Energetic, gregarious, with an easy smile and a quick sense of humor, Dominick was nevertheless strict in raising his children. Dave's mother, Lillian, from Polish stock, had been nearly crippled by a massive stroke when he was fifteen.

Dominick had taught his children that the key to success was sticking with the job. Dave, the oldest of the three, never forgot that. Still, he was a late bloomer. In high school his hobbies had been basketball, tennis, and girls, and even those he never pursued ardently. He was a loner as a teenager, one who avoided team sports, who was always the odd kid who didn't fit in. He had a small group of close friends rather than a large number of acquaintances. That was Dave's way. He was never considered cool. Most of his classes bored him and he earned only a mediocre B average. He never hung out with the really smart kids because he didn't have the grades. But he also didn't pal around with the youths who got in trouble.

Volonino scored high on his Scholastic Aptitude

Test during his senior year, high enough to be accepted into the prestigious Rensselaer Polytechnic Institute in Troy, New York. Though he had never been interested in school, he did want to do something with his life, and if he was going to go to college, he wanted it to be a good one. RPI had top programs in math and science, the two subjects that did interest him.

Dave was mechanically minded. His father was a fix-it guy as his son grew up. Together they rebuilt engines and transmissions, and, once, a television set— with Dave mostly just handing his dad the tools. They also tackled plumbing and electrical wiring chores around the house. Dominick loved to take things apart to inspect how they worked, and he always seemed able to fix them and put them back together.

Dave also wasn't intimidated by machines. He became a nuts-and-bolts enthusiast like his father. He started fiddling with the insides of stereos, rebuilding old ones whose parts had worn out. He began to admire the elegance of mechanical designs—the reason submarines, the most elegant of them all, would intrigue him later.

At Rensselaer, Volonino joined the Naval Reserve Officer Training Corps, which provided him with a scholarship. He worked part-time as a dishwasher to pay for room and board. He came from a family of sailors. His father had been a sonar operator aboard a diesel sub toward the end of the Korean War. The Volonino men were expected to join the Navy, even if for only one hitch.

Dave would be the first officer in the family, but he didn't grow up with a burning ambition to be a ship driver. He had attended college just after the Vietnam War ended, when the military was not popular. He had joined the Navy to pay for college. It was a business arrangement, plain and simple. He wasn't conservative or liberal politically. The military certainly wasn't his choice for a lifelong career. The Navy would pay for his education at RPI, and he agreed to serve for five years to pay it back. Then it was adios and off to the civilian world for a job. He wasn't a draft-dodging hippie, but whatever negative image his mind had of the military—and there was some, because he was against the Vietnam War—he set it aside when he signed his Navy contract. Only later, when he joined the submarine force and found that he loved what he was doing, did he realize the misperceptions he had harbored in his youth.

And only later did he realize that he had fallen into another family when he joined the NROTC at Rensselaer. Most of the school's students were nerds from wealthy families, and again, he didn't fit in with these brainy rich kids. The only students he had anything in common with were the ones in NROTC. They came from families of modest means like him. The other students, the snobby ones, called them "NROTC turkeys," but Volonino stuck with them.

He was never an impressive student in college, particularly in theoretical subjects. He eked out a 3.0 average, which was good but certainly not dazzling. Volonino didn't excel until later, when he began what amounted to graduate work learning the Navy's reactors. The Navy demanded it. The service was paying him and wanted its money's worth.

Dominick had talked to his son about serving on submarines as he had. Dave at first wanted no part of it. What person in his right mind would volunteer to be in one of those black sewer pipes?—which was about what subs had been when Dominick rode in them.

Submarines had had a checkered history. Aristotle, Herodotus, and Alexander the Great had all toyed with ideas for a submersible craft. Leonardo da Vinci is credited with the first design for a diving machine. By the 1600s, working submarines had been built, and during the American Revolutionary War, one had been tested in three attacks, all of which failed. During the Civil War, a crude, hand-cranked submarine built for the Confederacy by an Alabama sugar broker named Horace Hundley became the first underwater vessel to sink an enemy ship, and to lose its entire crew in the attack.

In World War I, the underwater wolves of both sides sank millions of tons of shipping. Even so, the admirals regarded them as little more than smart mines. World War I subs spent most of the time on the surface. For their crews, life was unbearable: no privacy, one bathroom for the some thirty aboard, no showers during long patrols. After ten hours submerged, oxygen dwindled and sub compartments filled with carbon dioxide, noxious gases from the engine, and the stench of rotting food. A revolution in sea warfare had begun, however. Homing in on their prey underwater, submarines were nearly impossible for surface ships to find and destroy—unless you drained the Atlantic Ocean first, as humorist Will Rogers once suggested.

By World War II, the technology of submarine and antisubmarine warfare had improved. Sonar to spot enemy vessels was developed, along with snorkels to suck in oxygen and vent gases for longer submergence. American submariners made up less than 2 percent of the Navy's manpower but were responsible

for more than half the enemy shipping destroyed—
although in the first year of the war almost one-third of
their commanders were fired because they were too
timid in battle.

The bolder skippers paid a price. One out of every
five submariners died in sub sinkings or accidents, the
highest casualty rate in the Navy.

Volonino had begun college thinking he wanted to
be a Navy pilot, but soon changed his mind. To get a
taste of what the service had to offer, NROTC midshipmen during their summer breaks spent about a
month each with a ship, a submarine, an aviation
squadron, and the Marines (who are part of the Navy
Department). Volonino found ship life boring and the
aviators seemed to him to be immature show-offs. The
Marines were impressive, but he couldn't see himself
as a ground-pounder. The submariners, on the other
hand, were a fraternity he soon became desperate to
join.

They were such quiet professionals, he marveled.
Their work could be as dangerous as a pilot's, but they
weren't daredevils. Submariners used their wits to
battle risks. Volonino was sent to the USS *Bluefish*,
a nuclear attack sub patrolling the Atlantic. He felt
overwhelmed at first. All around him were junior officers only three or four years older than he, who were
operating this complex machine and rattling off nautical terms in what sounded like a foreign language.
They didn't show off. But they exuded confidence as
masters of their technology. It blew him away. (He
never had that feeling when he was with the ship
drivers or the aviators or the Marines.) A submariner's
job seemed incomprehensible, and he would be expected to perform it after he graduated from college.

Volonino at first doubted he could do it. But he wanted to try.

The *Bluefish* crew didn't give him time to stand starry-eyed. They shoved a qualification card into his hand with a list of jobs to learn on the boat and told him to get busy. Volonino started studying the *Bluefish*'s trim and drain pumps, its hydraulics and power plant and electrical system. Another thing surprised him. The sub's sailors and officers made it their personal mission to help him learn. In other units, the midshipmen were considered pains in the asses that sailors had to baby-sit. But these submariners wanted him to be a part of their world from the moment he climbed through the hatch.

Volonino left the *Bluefish* convinced he wanted to be a part of this elite community. But first he had to get through Hyman G. Rickover.

A Polish Jew who had earned only respectable grades at the Naval Academy and had commanded nothing better than a minesweeper before he turned to engineering ships, Rickover was the father of the nuclear Navy. The first nuclear submarine Rickover built, the USS *Nautilus*, set sail in 1954. From then on, submarines powered by compact nuclear reactors could travel underwater at high speeds almost indefinitely, because they no longer needed an oxygen supply as diesel subs did. Abrasive, ruthless, and territorial, Rickover was unpopular with the Navy brass. He had been twice passed over for admiral before Congress finally ordered the service to promote him. But he built his own empire as director of the Naval Propulsion Program, refusing to allow any officer to serve in the nuclear organization unless he personally approved him.

He instilled a mind-set in his "nukes" that to err was

not human. In other words, it was possible to be perfect. By the time he was forced to retire in 1982 at the age of eighty-two, practically all of the Navy's subs were powered by reactors Rickover had engineered.

In 1978, Volonino, who had just turned twenty-one, paid his visit to Rickover, who by then was sixteen years past the service's normal retirement age. Volunteers for submarines must first receive an invitation from the Naval Propulsion Program to attend a screening at its Washington, D.C., headquarters, and not all volunteers were invited.

After two days of grueling written and oral exams, Volonino sat in an anteroom to Rickover's office, his brain wrung out, convinced by then that life would end for him if he was not accepted into the nuclear submarine program. He fidgeted in his seat, replaying the mistakes he knew he had made on the written tests and the stupid things he had said during the oral exam. He became clammy and tried to slow his breathing and will his pores not to sweat anymore. Stories were legion about Rickover playing mind games with midshipmen and probing for weaknesses until they left his office as quivering wrecks. The slightest faux pas could set the old admiral off and get a midshipman thrown out of the program before he even began it.

A stern-faced Navy captain in the anteroom finally called out Volonino's name, and his heart jumped through his throat. As Volonino stood erect before the doorway and smoothed out his uniform, the captain grabbed his arm and whispered last-minute instructions: "Don't extend your hand to shake his hand. Don't introduce yourself. The first words out of your mouth should be to answer his first question. The last words out of your mouth are your answer to his last question."

Volonino felt his throat tighten as the captain shoved him through the doorway. Rickover's office was spare but neat, with model ships scattered around on shelves. The admiral looked thin and frail behind a huge wooden desk. His head, with its snow-white hair, large ears, and wide eyes, poked out of a loose-fitting business suit. He seemed so small and harmless.

But then he spoke. Then all Volonino could see was that big head, the glaring eyes, and those white teeth bared like those of a German shepherd about to attack.

"Are you married?" Rickover finally said.

"No sir," Volonino answered, barely able to get the words out.

"Are you engaged?" Rickover asked.

"No sir," Volonino answered, becoming more mystified.

"Well, do you date anyone!" Rickover shouted angrily.

"Yeah," Volonino said, choking out the word.

"How many girls do you date?"

Volonino's brain was now scrambled. Why the hell was Rickover asking him about girls, and why was he getting angrier with each answer? His mind now raced to come up with an answer that wouldn't send the admiral into orbit again. Volonino knew that Rickover did not like having married men in his program. He wanted nothing to distract them from his reactors. Volonino was dating his future wife by then. Rickover was treading dangerously close to what Volonino suddenly realized was his weakness.

He tried quickly to collect his thoughts and to come up with a plausible number. One wasn't a good answer. More than two, and Rickover probably wouldn't believe him.

"Two," Volonino blurted out.

"Two?" Rickover bellowed. "Is that all? Just two?"

Rickover kept probing. Was Volonino contemplating marriage? he asked.

Volonino hemmed and hawed but left the admiral with the impression that he would not be walking to the altar soon, which technically was true.

Rickover finally tired of the jousting. He approved the application and ordered the young midshipman to leave. Volonino scurried out of the office, his uniform shirt soggy under his coat.

Now, twenty-one years later, he was the captain of one of Rickover's nuclear-powered submarines. Volonino checked his watch. It was 7:40 A.M., less than a half hour before the *Nebraska* was scheduled to push away from the pier.

"All right, the cigar-smoking light is now lit on the bridge," he announced and pulled out of his breast pocket a fat cigar.

Boyd radioed up from control. The sub was "ready to get under way." A third tug had pulled up to guide it out.

Ten minutes later, Captain Hunnicutt climbed up to the bridge so he could watch how this commander piloted out of the channel.

To break the ice, Volonino tried to show off his new toy, the SVS-1200 that electricians had installed with its colored map, but Hunnicutt interrupted him to lecture on the rules for its use.

"You understand that it can only be an auxiliary guide," Hunnicutt said stiffly.

"Yes, sir," Volonino answered, the smile wiped off his face. The intelligence he'd picked up on Hunnicutt

was correct. He wasn't a joker. All business. Freeland
and the other crewmen on the bridge grew quiet.

A crane hoisted up the gangplank off the sub's stern
and swung it around to the pier. Sailors on the sub unhooked dock lines from their cleats, and pier workers
pulled the ropes back. Hunnicutt began punching buttons on the SVS-1200 as if he was checking to see if
the crew had installed it properly, then flipped out a
steno pad to begin making notes.

It was now 8:05 A.M. on May 6, 1999. In Eastern Europe, NATO warplanes were in the middle of another
day of punishing air strikes against Serbian tanks and
armored personnel carriers in Kosovo. The month before, Russian President Boris Yeltsin had warned that
his country might be drawn into the conflict, which
could lead to "a world war." And now the USS *Nebraska* was setting sail for the eighteenth patrol of the
ship's life.

"Cast off all lines," Volonino ordered. The last thick
rope was unhitched from a deck cleat.

"Stand by to shift colors," he commanded in the next
breath. "Shift colors."

Chambers, who had been clutching the American
flag Hardee had given him, strung it up on a pole near
the radar mast, where it began to flutter in the breeze.
Freeland gave the ship's horn a long blast. It sounded
like a foghorn, and dockworkers covered their ears.
Sub bravado could be irritating.

2 • The Channel

Two of the three tugs, their lines now attached to the submarine's cleats, began gently pulling the *Nebraska* away from the pier and then nudging it around so its bow pointed southeast out the Kings Bay Entrance Channel.

When the sub was completely turned around, one tug sailed ahead of it as a guide. The second tug posted itself off the *Nebraska*'s starboard bow and the third trailed along the port side of the stern, each ready to nudge the long black vessel right or left if it sailed off the center track of the channel.

Volonino looked ahead, his eyes squinting as he tried to gauge if the sub was in fact steaming dead center down the channel. It seemed to be. The old harbor pilot standing on top of the sail agreed. But the bridge's differential GPS screen had the *Nebraska* slightly to the left.

"I guess we're okay," Volonino finally said, one eye cocked. "Maybe slightly to the left."

Brady radioed from control. His plotters also thought the sub was too far left and recommended it veer right.

Less than a half mile away from the wharf to which his boat had been tied, Volonino had Freeland order a course change. He knew from long experience that it was best to be nearly perfect as you began the trip down the Kings Bay Entrance Channel.

Only about 10 percent of the sub's time at sea was spent on the surface moving in and out of port. There were no ports in the world deep enough for the sub to leave the dock submerged, so it always had to sail out for a distance on the surface until it reached ocean waters.

But the small amount of time the *Nebraska* spent on the surface could be the most hazardous for Volonino. When he was submerged deep in the ocean and far from land, navigation became infinitely easier. The seas were still the loneliest places on earth. The *Nebraska* could stay far away from the few subs that patrolled them and sail well under the deepest-draft ships. But the submarine was out of its element traveling on the surface. Because so little of it stuck out of the water, watchmen on distant ships had difficulty spotting the sub, even through their binoculars. The *Nebraska* had backups to backups in its navigation system, but computers could fail, turn orders could be misinterpreted, lefts could be confused with rights, and the sub could be on a collision course or run aground within seconds. A captain's career could be ruined if he wasn't careful.

As far as sailing in and out of a port goes, Kings Bay is a lousy location for a Trident submarine base. Its only saving grace is that the other ports where the Navy docks its submarines are even worse. The problem at Kings Bay is the thirteen miles of narrow channel the Tridents have to sail through in order to reach deeper water in the Atlantic Ocean.

It is like driving along a skinny, crooked road. To

travel this channel, ten precise turns have to be made. Many of the legs are less than three hundred yards long, so as soon as the helmsman makes one turn another is upon him. The sub first has to head southeast out of the Kings Bay Entrance Channel, then make a series of right turns to sail south down the Cumberland Sound, with the marshy Georgia coast to the west and Drum Point Island and Cumberland Island to the east. As the sub nears the southern mouth of Cumberland Sound, where the St. Marys River and Jolly River pour in from the east, the channel doglegs to the left to the St. Marys Entrance. The St. Marys Entrance is the waterway between Cumberland Island on the north and Amelia Island to the south. From there, the channel stretches out another eight miles east into Atlantic waters that are finally deep enough for a sub to sail anywhere on the surface.

Until then, the *Nebraska* would have practically no maneuver room. The bottom of its hull sank thirty-eight feet into the water. The channel for the sub to sail through had been dredged to forty-seven feet at the most, which left no more than nine feet of clearance between hull's bottom and the seabed. When the sub increased speed, it also tended to squat even deeper into the water, so there would be only a few feet between the boat and the channel's bottom.

What's more, the channel is thin. Broad stretches of it are only a little wider than the length of the sub. There is no room to turn around in it. A sub off course could drift outside the channel within seconds and strike bottom. Or just as bad, if it got turned crossways in the channel, the sub could never line itself up again properly. Tugs would have to come to the rescue.

The problems with the cramped traveling are compounded by the fact that on the surface, a Trident can be a nightmare to turn. Underwater, with its rudder and fins all working hydrodynamically, it is nimble. But on the surface, it is cumbersome, because its hull is rounded at the bottom and only half of its rudder sticks into the water to turn it. Crosscurrents in the channel can shove a submarine in different directions. The vessel has to keep up a speed of at least eight knots so the rudder can bite into the water and force the ship to turn. At three knots, it can take a half hour for a sub to make a full turn. Any slower and it becomes uncontrollable.

The *Nebraska* now picked up speed as it drifted past the two explosive handling wharves on its right, where the D-5 missiles are loaded into the Trident tubes. A frothy wake spread out from its bow like a white blanket. One of the three tugs peeled off, leaving the other two to guide the sub out. Sailors on the deck now flipped upside down the cleats that had held mooring lines and began beating the hatch covers over them with rubber mallets. The hatch covers over the cleats had to be locked in place snugly. If they rattled or created turbulence over the deck's smooth surface when the sub was submerged, that slight noise might be picked up by an enemy's sonar. The beating mallets sounded like timpani playing. A sailor could tell if the hatch covers weren't locked firmly in place by the change in pitch or the vibration he felt from the mallet's handle.

A sailboat drifted across the channel about six hundred feet in front of the *Nebraska*. During the summer months, Trident skippers are constantly harassed by pleasure craft that sail around their subs. The curious ones often get dangerously close, not realizing that the giant black vessels can't turn in the channel to avoid hitting them. Some women in speedboats even like to charge up to the subs and yank off their bikini tops to give the topside crew a show. But today, a harbor security boat that patrolled alongside the *Nebraska* raced ahead to shoo this sailboat out of the channel.

"Captain, you need the tugs any longer?" the harbor pilot standing on the top of the sail finally yelled down to Volonino.

"No, sir," Volonino answered. He could handle the rest of the channel voyage on his own. The two remaining tugs peeled off as the *Nebraska* made a slight right out of the Kings Bay Entrance Channel where the Cumberland Sound joins in from the east.

Al Brady, the navigator, phoned from the control room. The sub was slightly off center. He recommended a course of 150 degrees for part of this 1,450-yard leg. Freeland glanced back at Volonino for approval.

Volonino eyeballed the channel. "Yes, I concur," he said quickly. "One-five-oh is a good course." Freeland relayed the command on his microphone.

"Passing buoy five-one starboard," Chambers shouted from his perch, as the sub sailed by a buoy bobbing in the water to the right; it was green, with a white 51 painted on it. Buoys are chained to the sea bottom to mark the boundaries of where the channel has been dredged. Green buoys are placed along the western side of the channel, red buoys on the eastern side. Volonino knew that if his sub strayed too many yards on the other side of the buoy it would strike bottom at the channel's edge.

Volonino checked his SVS-1200 screen to compare the sub's position on it with what he saw in the channel. Hunnicutt had so far said little on bridge, making notes occasionally on his steno pad. On its right the sub now passed the magnetic silencing facility, a giant series of some twenty electrically charged metal arches built over the water. They look like an Erector set shell for a huge barn. Tridents pass through them to remove any permanent magnetism their hulls have picked up during a voyage. That will reduce the magnetic signature a submarine gives off, which might enable enemy vessels to identify it.

The *Nebraska* was nearing its next way point, where it would have to make another right turn. In the control room below the bridge, the piloting party busied itself with calculations of the sub's position and heading.

Quentin Albea, a twenty-four-year-old yeoman, sat behind the right wheel in front of the ship's control panel. He watched intently the analog and digital direction indicators in front of him, turning the wheel ever so slightly to the right or left to keep the sub steaming on the 150-degree heading. One of the sub's more experienced helmsmen, Albea controlled the stern rudder, which steered the boat right or left. He planned to take part-time classes in computer programming at Valdosta State University when the sub returned from this patrol.

Boyd, the ship's executive officer, roamed between the chart tables and the radar screens to their left. Once in a while he would hop up to the conn and peer through the number two periscope to scan the outside. He had two jobs now: monitoring the piloting party to make sure it kept the sub headed in the right direction and watching out for other vessels to keep from colliding with them.

Next to Boyd was Petty Officer First Class Ed Stammer, who pressed his face to periscope number one's eyepiece and gripped its two handles, which looked like a crossbar. A hulking Texan, Stammer slowly swiveled the scope around in a 360-degree turn. The sailors called this "dancing with a flat-chested woman." By twisting the grip on the periscope's right handle, Stammer could increase the lens magnification from low power to high power. By twisting the left handle, he could tilt the lens up so he could look above to make sure no plane or helicopter was overhead. Nuclear attack subs that hunt enemy vessels have even fancier gadgets on their periscopes, which can electronically transmit the image an operator sees in his eyepiece to a television and VCR. This allows the captain to watch the prey and analyze the picture on a video replay.

Stammer was now searching outside for buoys and towers protruding from the water and land markers called "ranges," all of which he could line up in the scope's sights to fix his sub's position in the channel. He had been up and down the channel for more than ten years, long enough to develop a seaman's eye for calculating the distances and angles from the ship to the ranges he saw in his periscope's crosshairs.

Brady and Standley, the piloting party's navigator and senior chief, stood over the chart at the primary plot table, where Suzor slid a parallel motion protractor over the nautical map, marking straight lines and checking distances with his divider. There were two plotting tables in control, a primary and secondary, with each making independent calculations of the sub's position and recommendations for its heading. All of this was part of the redundancy in navigation that existed aboard ship. One backed up the other. Brady spent most of his time stooped over the primary plot, which usually marked the sub's position on the channel chart based on visual fixes Stammer called out from the periscope and signals received from global positioning satellites. He left one of his most skilled petty officers, thirty-year-old Chris Cornelius, to supervise the secondary plot, which fixed the sub's position on the chart based on latitude and longitude readings from the GPS and ship's radar.

The key to navigating is plotting where you were going to be rather than where you are. By the time Suzor and Cornelius managed to fix the *Nebraska*'s position on their charts, the sub had already moved ahead fifty yards. Their pinpointing was always thirty seconds or more in the past, ancient history as far as navigation was concerned. To make up for this, the two sailors drew "dead reckoning" lines out from the fixes they had plotted on their charts to predict where the sub was in the channel at the moment or should be in the future. It was as much art as science.

Almost every other minute, a round of reports on the sub's position now flooded into Brady's brain—from Stammer's sightings through the periscope, from Suzor and Cornelius, from the GPS and radar operators, from the bridge's radio traffic, from the fathometer operator, who shot a sound pulse through the water to measure the hull's distance from the channel's bottom. In rapid fire, he mentally processed the information and spit out recommendations to Volonino on the turns the ship should make.

Brady pressed his small microphone to his lips. The sub had almost reached the point on his charts where it needed to make another right turn so it could line up on the next leg of the channel on a southwest course of 171 degrees.

"Bridge, navigator," Brady phoned Freeland on the bridge shortly before he thought the *Nebraska* should veer right. "Five hundred yards to the turn, counting down the turn."

"Navigator, bridge, aye," Freeland answered.

The turn had to be timed precisely.

"Three hundred yards to the turn," Brady phoned Freeland. "Two hundred yards to the turn. One hundred yards to the turn. Stand by to mark the turn. . . . Bridge, navigator, mark the turn."

"Mark the turn, bridge aye," Freeland answered quickly and turned to face Volonino.

Volonino nodded.

"Helm, bridge. Right fifteen degrees rudder, steady course one-seven-one," Freeland said over his 7-MC phone to Albea, the helmsman in the control room.

"Right fifteen degrees rudder, steady course one-seven-one. Bridge, helm, aye," Albea radioed back and turned his wheel clockwise.

The *Nebraska* slowly swerved to its new heading, a frothy white wake now trailing it for five hundred yards. It passed Drum Point Island on the left, then the marshes that Mill Creek wound through on the right. The timpani concert from the pounding of hatch covers had finally stopped.

Chambers called out the numbers as the sub passed buoy 47, then 46, 45, and 44.

Volonino glanced down at his SVS-1200. Its reading seemed dead on for the position at which he fixed the sub.

A hot sun beat down as Cumberland Island next came into view on the left.

"This is pretty sweet," Volonino said, admiring the computer screen and the day.

But not for long.

It happened almost in an instant.

The *Nebraska* had just passed buoy 42 on the left and Freeland had begun turning the ship right so that it would be on a 184-degree heading, almost directly south down the Cumberland Sound. But the harbor pilot noticed that the turn was being taken too slowly and warned Volonino that he better make it tighter or the sub would end up left of center on the next leg.

Volonino agreed. "Increase your rudder to right full," he told Freeland. "Steady on course one-eight-four."

Freeland keyed his microphone and repeated the order to the control room. Then came the breakdown.

The first part of Freeland's order down to Albea at the helm—"Right full rudder"—came in loud and clear over the speaker in control. Albea immediately turned his wheel clockwise to begin turning the submarine to the right. But Freeland's phone circuit suddenly went dead, and the second part of his order—"steady course one-eight-four"—never reached Albea.

That was a critical piece of information that was now missing. "Steady course one-eight-four" would have told Albea that he was to stay in the right turn only until 174 flashed on his direction indicators, ten degrees short of 184 degrees. At that point he would quickly turn the rudder to the left to counteract the rightward momentum, then center his wheel so the submarine would end up on a southerly course of 184 degrees. Because Albea was steering blind in the control room, he had to follow without question the turn orders that came to him. And this order told him to turn to the right—and keep turning.

The *Nebraska* slowly began making its barnyard turn to the right.

The right turn continued and continued. And continued.

Instead of heading south at a course of 184 degrees, the *Nebraska* was now sweeping toward the right side of the channel and barreling head-on toward buoy number 41.

In the next second, there was an explosion of orders and chaos.

Volonino shouted the 184-degree course call again, but Freeland was repeating the command into a dead mike. The *Nebraska* was now on a collision course with buoy number 41. In less than two minutes, the huge sub would ram it.

On the *Nebraska*'s second level, Chad Thorson was staring at the computer screen in his cramped stateroom when he was jolted practically out of his seat from the shouts above him. Pandemonium had erupted in the control room on the next level up. The navigation team realized the sub was dangerously off course. Boyd was yelling at the top of his lungs.

Martin, who was standing next to Freeland on the bridge and manning the backup phone, had relayed in the next second the 184-degree course order to a sailor on the other end of the line, who was in the control center, sitting to Albea's left. But in the loud ruckus around him, Albea never heard the sailor when he passed on the course order from Martin.

Volonino had a third way to communicate with the control center: a handheld radio with him on the bridge, which he could use as an emergency backup to speak directly to Boyd. When he had tested the radio just minutes earlier, it worked fine. But in those intervening minutes, a sailor in control had switched channels on the center's radio to talk to the shore command—and had forgotten to switch the radio back to the frequency on which Volonino was speaking. So his executive officer's radio receiver was silent as Volonino barked out orders to try to reverse the right run.

Murphy's Law had paralyzed the control room. Boyd was new to the sub, and the rest of the navigation team were rusty after four months of shore duty. Even so, this was not their finest moment.

"Watch it now . . ." the harbor pilot standing on the sail began, but never had a chance to finish his sentence.

Volonino, in that same instant, realized his turn orders had been horribly garbled down below.

"BACK EMERGENCY!" he screamed. "ALL BACK EMERGENCY!" Volonino was slamming on the brakes by kicking the *Nebraska*'s steam turbines into reverse. It was the only way to stop the sub from hitting the green buoy or, even worse, sailing past it and running aground on the right bank of the channel.

This time, the directions finally got through to the engine room. Martin repeated it on his phone, which still worked. And Volonino grabbed Freeland's shoulder and ordered him to relay the command over the sub's general phone circuit for routine announcements, the 1-MC, which was also working.

Todd Jenkins had already begun to push up slightly from his chair in maneuvering. He had nervously sensed that something wasn't quite right from the broken transmissions he had just heard over the intercom that normally conveyed orders from the bridge.

Maneuvering was a restricted room the size of a

walk-in closet located just behind the nuclear reactor compartment on the upper level of the sub. To get to it, seamen with the proper security clearances had to open a hatch, which remained closed at all times, and walk down a long hall above the reactor compartment on the starboard side. No civilians, and very few Navy men, are allowed into the back end of a Trident submarine. The service jealously guards its reactor secrets. Inside the air-conditioned room, three technicians sat before panels crammed with rows of indicators and knobs to operate the reactor and the sub's electrical plant. Also in front of them were two throttles that controlled the speed of the sub's propeller. Behind the technicians, an engineering officer of the watch sat at a desk whose lockers underneath were stuffed with secret manuals for running the plants.

Jenkins was the throttleman who sat in the far left seat facing the panels. A soft-spoken African-American who had spent two years as a chemistry major at Fisk University before joining the Navy, he had been nervous at first about sitting in that seat during surface maneuvers. It was a big responsibility. Any job you had to be interviewed for by Commander Volonino must be important, Jenkins realized. The throttleman who stood watch when the sub sailed in and out of the channel was handpicked by the skipper. There were other throttlemen aboard who were senior to Jenkins, but Volonino considered Jenkins his most reliable one. For the quick speed changes required during channel operations, the CO wanted a sailor who read the panel indicators quickly and reacted to orders from the bridge instantly.

Tony Holmes, a senior electrician who normally sat

in the far right seat facing the panels, now stood behind Jenkins with a sound-powered phone strapped around his neck.

The back emergency order crackled in Holmes's earpiece.

"Back emergency!" Holmes shouted at Jenkins.

The room turned deathly quiet. His heart pounding, Jenkins reached up and twisted the knob on a large box which sounded a bell to signal the control room that he had received the order. Then he grabbed the outer wheel in front of him, which was almost two feet in diameter. That was the ahead throttle, which made the sub go forward. With all his might, Jenkins quickly spun the heavy wheel clockwise about ten revolutions, which eventually brought the sub's propeller to a dead stop. Next he grabbed the astern throttle, a wheel set inside the ahead throttle that was a little smaller than an automobile steering wheel, and began spinning it counterclockwise for a half-dozen revolutions. That caused the propeller to whirl around in the opposite direction to pull the sub back.

Lieutenant Commander Harry Ganteaume, who had been in the upper level of the engine room, dashed into maneuvering as soon as he heard the back emergency order. Ganteaume was the engineering officer. The reactor and practically everything else in the back of the sub that powered the *Nebraska* was his preserve. By the sound of the electrician's voice, Ganteaume knew this wasn't a training drill. The ship was facing a major casualty if the turbine engines didn't stop its forward movement.

"Easy, easy," Ganteaume said softly as Jenkins spun the astern throttle. He didn't want the excited throttleman to exceed 100 percent steam power on the reverse propeller speed, the sub's limit.

Almost immediately, Ganteaume and the four maneuvering room operators felt the *Nebraska* going into reverse. Water now rushed forward over the stern as a result of the propeller's spinning the other way. A scary rumbling sound swept over the entire engine room. The deck in maneuvering began rattling like an earthquake. Dust came out of the overhead, and books fell off the desk behind the operators.

On the bridge, Volonino looked quickly to the back of the sub. He could feel his ship shudder as well. White waves billowed up at the stern as the propeller shifted into reverse.

But the 18,750-ton vessel couldn't stop on a dime, not hardly. At the speed the *Nebraska* had been traveling forward, it would take more than a thousand feet for the propeller spinning in reverse to bring this monster to a dead halt. Volonino guessed he had about a thousand feet between him and buoy 41. But he wasn't absolutely sure.

As the propeller fought to pull the sub in the opposite direction, the *Nebraska* came closer and closer to the green buoy, the tip of its bow pointed directly at the huge metal float bobbing in the water. If the sub slowed down enough so its bow just tapped the buoy, that would be a black mark on Volonino's record but an accident he could survive. If the *Nebraska* bowled over the buoy and, heaven help him, ran aground on the side of the channel, that would probably end his naval career. The service could be brutal to captains who damaged their vessels.

"Come on, ship, come on!" Volonino growled.

He had ordered the rudder centered at the stern. There was no use turning it now. The sub wasn't moving fast enough for the rudder to deflect the bow from the buoy.

He had perched himself up on top of the bridge, sitting on its rim, for a better view of the distance between the *Nebraska*'s bow and buoy 41.

The distance was becoming shorter and shorter: Five hundred feet, four hundred feet, three hundred feet. The sub continued to shake. Volonino could feel its forward speed decrease as the propeller strained to bring it to a halt. But the giant *Nebraska* kept lumbering on as if it had a mind of its own, intent on slamming into the buoy.

"Come on, ship," Volonino growled again, in an even louder voice. "Come on!"

Ensign Chesney, standing in one of the lookout stations behind Volonino, stared bug-eyed at the buoy getting closer and closer, his mouth half open. He might be seeing the first collision of his naval career.

The *Nebraska* was now one hundred feet from the buoy. Then ninety feet. Then eighty feet. Then seventy feet.

The sub continued to move forward, but at a slower rate. Volonino was now pretty sure the *Nebraska* wouldn't hit the buoy, although he didn't know how close it would get. Tridents are incredibly sluggish above water, but they are predictable. When Volonino first came aboard, he had taken his sub out on the open ocean and tooled around—slamming on its brakes, scratching off from a dead stop, turning this way and that—to get a feel for what the ship could and couldn't do. Over the past two years, he had brought green officers like Chesney up to the bridge and had them watch how fast the submarine could stop so they would know its capabilities in an emergency.

Forty feet. Thirty feet. No one said a word on the bridge. Nothing more could be done at this point. The *Nebraska* had slowed to a crawl, but it was still crawling forward. Waves slapped the side of the hull and white foam sloshed in the propeller's wake.

Twenty feet.

The *Nebraska* finally came to a halt. Everyone on the bridge let out a collective breath.

"All back full," Volonino ordered, the command to slow the sub's reverse speed. With the *Nebraska* stopped, he now didn't want it backing up too fast so its stern shot across the other side of the channel.

Freeland's communications circuit was still dead. Martin had to relay the order on the backup system.

Volonino exploded from the tension on the bridge. "Get the comms fixed! NOW!" he shouted furiously at Freeland and Martin.

Martin hurriedly passed the order over his phone to have a maintenance technician hustle up to the bridge and fix Freeland's set.

Throughout the ordeal, Captain Hunnicutt had remained silent and calm. Volonino was impressed. Hunnicutt had followed protocol. Volonino was responsible for the ship, and it would have been bad form for Hunnicutt to step in and backseat-drive. Besides, Hunnicutt wouldn't have done anything differently with the emergency if he had been in charge. Tridents routinely ran drills where they trained on how to deal with rudders being jammed in one position or radios breaking down on the bridge or subs having to be stopped before they hit something. The real emergency Volonino had just faced was one sub captains simulated many times.

And Volonino had responded exactly as he should have: shifting to his backup communications and ordering back emergency. It was a "nonevent" as far as Hunnicutt was concerned.

Hunnicutt seemed to be a regular guy, Volonino thought. In fact, though he could have a hard edge when he needed to get things done, Hunnicutt turned out to be somewhat folksy. A Georgia native, he had almost twenty-four years in the Navy. Rickover had thrown Hunnicutt out of his office twice because he didn't like the answers he gave each time during his interview. On the third visit, the admiral had finally approved the trembling ROTC midshipman for the nuclear program.

When the *Nebraska* finally came to a halt, Hunnicutt did have one editorial comment to make.

"Okay, now you've stopped the ship," he piped up. "But where the crew can screw it up is in the recovery stage."

Indeed, most skippers did well in stopping their boats in time, Hunnicutt knew. The bigger headache was getting the huge submarine lined back up in the channel so it could steam forward.

Volonino knew as well that halting the boat was the least of his problems. He was now sitting practically dead in the water, with his sub just beginning to back up, but cockeyed in one of the narrowest parts of the channel. A stiff crosscurrent was coming in that could throw his ship off course even more. With the sub standing still, its rudder was useless. If the current pushed him sideways any more, he'd end up perpendicular in the channel and practically paralyzed.

"Captain, I'd recommend right full rudder," the harbor pilot said in a southern drawl. More recommendations came up from the control room on how to line the sub back up. It seemed that every crew member now had an opinion about how to get Volonino out of his fix.

Volonino tuned out the others. He had learned over the past two years that only one man's opinion counted on this ship. His. Volonino, in fact, had more experience than any of his crew in handling the sub. He had learned to trust his instincts and to go with what he thought was the right thing to do. He would drive the submarine from the bridge by himself.

"I'm going to back it into the channel," he said firmly. "And I'm going to go right full rudder." Volonino planned to move the rudder right as he backed up, to twist the sub so its bow would shift left and be pointed down the center of the channel again. It was a tricky maneuver, but the tone of his voice left no one on the bridge in doubt that this was how they were going to get out of this jam. No questions asked.

"Back two-thirds," Volonino snapped. He was angry and tense. The order slowed the sub's propeller once more as the Trident began moving backward quicker.

Slowly the *Nebraska* began twisting, as he'd known it would. The bow began shifting to the left, aligning itself along the middle of the channel.

When the sub was finally as near to the channel's center as he could get it, Volonino ordered, "All ahead standard, left full rudder." The propeller began spinning in the opposite direction and accelerating the sub forward, past buoy 41 to the right. With a little speed on the sub, the rudder began to bite in the water and turn the vessel when Volonino commanded.

"Shift your rudder, steady course one-eight-four," he ordered as the Trident regained the correct heading.

"All right, back in the groove," Volonino finally said,

letting out a breath, then added as an afterthought: "That's the first time that's ever happened to me."

The *Nebraska* continued steaming south. The St. Marys River emptied into the Cumberland Sound on the right, its current pushing the submarine slightly off course. On the left, wild horses grazed along the shore of Cumberland Sound.

The sub began the series of left turns for the dogleg around the southern tip of Cumberland Island. Hunnicutt turned his binoculars toward Fort Clinch, a Union fortress during the Civil War built on the northern shore of Amelia Island to the sub's right.

"Hey, Captain," Hunnicutt said to Volonino, putting down his binoculars. "Do the letters GBR mean anything?"

"Go Big Red," Volonino answered, and brought his binoculars up to his eyes to scan the Amelia coast.

On the beach of Fort Clinch State Park stood Chesney's wife, Heather, and Melissa Bush, who was the wife of Lieutenant Junior Grade Dave Bush, the sub's main propulsion assistant. They held up a white bedsheet with GBR painted on it in red. Wives always stood on that beach when the *Nebraska* set out for patrol.

"Ten-second test of the whistle," Volonino ordered with a smile.

Freeland gave it a long noisy blast.

Volonino looked through his binoculars again. Heather and Melissa were waving.

Volonino waved back. So did Chesney behind him. He wouldn't be seeing his wife of just three months for a while. The *Nebraska* steamed out the St. Marys Entrance into the Atlantic Ocean, perfectly centered as it passed buoy 22 on the left and 23 on the right. Volonino ordered the boat to speed up to thirteen knots so the crosscurrent from the rock jetties to the north would have less of an effect on the sub's heading. A stiff Atlantic wind whipped back the American flag above the bridge.

The *Nebraska* was now on a straight eight-mile leg of the channel out to the Atlantic. Volonino could relax some. He leaned over and cupped his hands around another cigar to light it. The tip of the stogie glowing red, he leaned his head back and blew out the sweet smoke.

"I figure that's my excitement for the whole patrol," he said to Hunnicutt with a chuckle, recalling his near collision with the buoy. "I don't need any more."

"Yeah, it's like sex," Hunnicutt answered, smiling.

"How's that?" Volonino asked.

"Real excitement. Then all's quiet."

3 • A New Life

The *Nebraska* spent the rest of the morning conducting training drills in the open ocean about fifty miles off the Atlantic Coast. The drills were endless. Practically every day the sub was on patrol for the next eighty-six, the crew would practice fighting mock battles or responding to emergencies. Submariners on board could sit for hours dreaming up unusual emergencies they might face at sea, then concoct drills to test how they would deal with them. The sub would not dive until later in the evening, so the crewmen now began a series of drills for surface operations. They tested how they would react if the sub was in a channel as narrow and shallow as the one they had just left with the periscopes broken, or with the signal lost from global positioning satellites, or with the rudder jammed fifteen degrees to the left. They simulated sailing through thick fog with a tanker nearby in danger of colliding with the sub. The *Nebraska*'s radio shack and Submarine Group Ten headquarters back at Kings Bay began exchanging code-word messages on mock emergencies to test the radio circuits. A cardboard box wrapped with duct tape (the crew nicknamed it "Oscar") was thrown off the ship to simulate a sailor topside who had slipped on the side of the rounded hull

and fallen into the water. The crew practiced driving
the sub back to pick up the man overboard.

Shortly after noon, Volonino climbed down from the
bridge and convened a meeting of the navigation team
in the navigation center behind the control room. This
was a postmortem on their near disaster with buoy 41.
It was a grim gathering. Communications circuits did
fail, nobody could be blamed for that, but too many
mistakes had been made in responding to the emergency, and Volonino now spent an unpleasant hour dis-
secting the foul-ups with his men.

Mark Nowalk strolled by the entrance to the navigation center, saw that the surgery without anesthesia had
begun, and quickly walked away. There were advantages to being an ensign and too junior yet for the surface navigation team. He was glad he hadn't been part
of that debacle. He hustled back to the missile control
center one level below, where he felt far more at ease.

Coming aboard the *Nebraska* had been a homecoming of sorts for Nowalk. It was like revisiting an old
house you grew up in. The first thing he saw climbing
down the ladder through the missile compartment
hatch four weeks ago had been a set of weights and a
workout bench on the upper level. He had spent the
four years he had been on the *Nebraska* before just
begging for that. Yeah, it seemed like a lot more attention was being paid now to making the place more hab-
itable for the crew. Weight sets and exercise machines
to keep them fit. Glass cases everywhere, even over
water fountains, that displayed University of Nebraska
memorabilia. Dolby Digital speakers for the sound
system in the crew's mess. A coffeemaker that made
not just the brown sludge he'd been used to but Colom-
bian coffee no less.

Every time he passed it, Nowalk couldn't help but steal a glance at the heavy brass commissioning plaque that hung on the bulkhead outside the galley. There it was, way down a long column of names on the plaque: MT2 M. A. NOWALK. Missile technician second class.

Nowalk wasn't new to submarining or to the *Nebraska*. At twenty-eight years old, he enjoyed two distinctions: being a "mustang" and a "plank owner." "Mustang" was the nickname given to officers who had first served as sailors. "Plank owner" meant that Nowalk had been part of *Nebraska*'s first crew when it was commissioned.

Nowalk had spent his high school years in Pennsylvania thinking he wanted to be a chef. At seventeen, he had worked as an apprentice in an elegant restaurant just north in Binghamton, New York. But the movie *Top Gun,* about daring Navy pilots, had just hit the theaters, and Nowalk decided that was how he wanted to see the world, aboard an aircraft carrier.

But the only world the Navy showed him was the one inside a submarine under the ocean. In 1988, the service packed him off to Navy boot camp in Great Lakes, Illinois, then back east to submarine school in Groton, Connecticut, where he studied to be a missile technician. After school, he crewed aboard a sub and served in a shore post from 1989 until 1992. Then he was transferred back to Groton, where he joined his second sub, the USS *Nebraska*. It was sitting on blocks at the General Dynamics Electric Boat Division plant with giant holes cut in its hull. They were the openings for the tons of mechanical and electronic equipment being loaded into the new vessel.

The *Nebraska* was the second U.S. Navy ship to be

named in honor of the thirty-seventh state of the Union. The first USS *Nebraska* had been a battleship commissioned in 1907; it transported troops and protected convoys during World War I. It was retired from the fleet in 1920. Nowalk spent his first year and a half on the new *Nebraska* installing and testing its ballistic missile system. Every weld on the twenty-four giant missile tubes had to be checked by hand. Finally, on August 15, 1992, Nowalk stood at attention on the sub's deck in his gleaming white sailor suit (he was a petty officer second class by then) and watched as Pat Exon, the wife of the state's senior senator, shattered a bottle of champagne on a railing next to the sub to christen it. (A Trident's hull is too round and sunk too deep in the water for a bottle to be broken against it.)

After the first two cruises on the *Nebraska*, Nowalk was transferred to Kings Bay, where for three years he instructed seamen on the Trident D-5 missile system. Many of the enlisted men on the sub now had been his students. But Nowalk wanted to be an officer, for the better pay and greater respect you got with the job. He was finally commissioned on October 1, 1998, a slightly overaged ensign by then.

It felt strange coming back to the *Nebraska* with gold bars on his collar. The officers' study on the sub's second level had been a room he visited before only if he was invited. Now he could stroll into it anytime he wanted to read the ship's manuals. His first dinner in the wardroom, where just the officers ate, was an experience. As a petty officer he had queued up with the rest of the sub's enlisted men in the chow line and taken his meals cafeteria-style from the galley. Now he sat in a private dining room with linen on the tables and cloth napkins and fancy silverware. A mess steward served him his plate and asked him if he wanted salad or soup.

That first day back on the sub, Nowalk spent three hours just wandering around, catching up. It was a marvelously self-contained world. The temperature kept at almost a constant sixty-eight degrees. Fresh water distilled from outside seawater. Oxygen supplied by machines converting seawater, or by air banks that stored it. Foul air cleansed and recycled by carbon dioxide scrubbers and charcoal filters. The ship's engineers could regulate the pure oxygen in the air mix, pump it up if it grew stale, or lower it to reduce combustion dangers. White lights for day and red lights for night hung in many rooms to help biological clocks keep the change in hours straight.

All of the *Nebraska*'s rooms and passageways were surrounded by a spaghetti of dials and switches and color-coded pipes and outlets. Dark green designated the sub's trim and drain system. Blue valves were for fresh water, orange for hydraulic fluids. Emergency phones were marked with orange stripes. Green, gray, and tan designated different types of air valves. Outlets, with red arrows pointing toward them, stuck out everywhere, to which crewmen could attach rubber tubes and inhale air from breathing masks in emergencies such as fires (they called it "sucking rubber").

Nowalk worried about how he would be received when he made his first rounds in the sub, particularly by the sailors he knew from his enlisted days as an instructor. Just several months ago he was "Mark," who had gone out to drink beer with them and played in their softball league. Now he was an ensign, which required a "sir" after every sentence from them. Nowalk dreaded the first time he walked down a passageway and one of his old sea buddies might say "Hi, Mark" instead of "Hello, Mr. Nowalk," as protocol now demanded. He would have to pull the man aside and set him straight, which he dreaded doing. Nowalk, tall and thin, had a gentle manner. The service had rules against officers and enlisted men fraternizing, but he never wanted to be one of those martinets as an officer.

Fortunately, all his old friends adapted well to his new rank. They recognized without prompting that he was an officer now. Only the weps, Lieutenant Freeland, who was now his boss, slipped up and called him Petty Officer Nowalk. Embarrassed, Freeland apologized.

"Don't worry about it," Nowalk said good-naturedly. "It's taken me a while to get used to it, too."

Brent Kinman stood rocking from the heels to the balls of his feet, patting his beefy fist into his beefy hand. If there was a minute when Kinman was awake that he wasn't in constant motion or excited about something, no one on the crew could remember it. Lieutenant Brent Kinman was a perpetual energy machine, like the sub's reactor.

He was the *Nebraska*'s drill coordinator for this patrol, which became almost a full-time job because the sub ran so many training drills. Kinman and his assistant, Chief Shawn Brown, scurried about the ship at all hours of the day and night setting up drills, then monitoring how the crew performed during them. It became a thankless task. The crew hated the drill monitors because they were always waking them up in the middle of the night with simulated emergencies. And when the crew performed badly on one of the exercises, the senior officers invariably blamed the drill coordinator. Now he and Brown, wearing red drill team ball caps and headphones with pen mikes so they could communicate with each other, watched from two corners of the command and control center to observe how the room dealt with the final drill for the afternoon set.

"The crew is worn out," Kinman whispered almost to himself. Certainly he was, and the patrol had barely started. The grueling refit and the morning drive out of the channel had exhausted everyone. Kinman nevertheless began fidgeting as he always did when a drill began.

"Romeo eleven is maneuvering on us," Al Brady announced on the radio circuit to the bridge. "He could be an AGI."

"Inform the XO," Hardee radioed back from the bridge, where he had taken over as officer of the deck, with Volonino, who had climbed back up the sail.

"Bridge, control, aye," Brady answered and reached for another phone to summon Lieutenant Commander Boyd.

For the past half hour, control had simulated tracking two unknown vessels from the surface. The first ship, designated romeo ten, had finally sailed away and didn't appear to be a threat. But the second ship, designated romeo eleven, was steaming from the south toward the *Nebraska* and getting closer. Brady now guessed the vessel was an AGI, which stood for auxiliary gathering intelligence.

During the Cold War, auxiliary gathering intelligence ships from the Soviet Union would loiter outside the St. Marys Entrance waiting for the Tridents to emerge into the Atlantic. On the outside, most of the surveillance vessels looked like harmless trawlers, but inside they were crammed with sophisticated electronic equipment to vacuum any sound, radar, or radio wave the Tridents emitted. The signals were analyzed aboard the trawler or fed to Soviet attack subs lurking underwater that tried to develop electronic profiles of each American sub so they could be recognized and tracked. Sometimes the bolder Soviet trawlers rushed up to the Tridents to try to force them off track in order to gauge their turning capabilities.

It could be a hassle running the AGI gauntlet so the Tridents could sail into the Atlantic to hide. Over the years, submarine captains had developed a number of evasion tactics to lose the pesky trawlers. One of the best ways to foil the electronic eavesdropping, the skippers had found, was simply to raise a racket as they sailed off from the coast. Air blowers and machines on the submarine were cranked up full blast to mask any sounds that normally came out of the subs, so the trawlers could never get an accurate acoustical signature for each boat when it operated normally.

Nowadays, a Russian surveillance trawler rarely parked itself along the American coast. But the *Nebraska* still trained in case a swarm of them returned.

"Romeo eleven bears one-two-eight, range seven thousand yards," announced Freeland, who had hopped up to the conn as the contact coordinator. Freeland's job now was to track the AGI ship based on readings from the radar. By his calculations, the trawler would be within five hundred yards of the sub in sixteen minutes if both vessels stayed on their courses.

Hardee radioed the engine room to crank up the sub's speed and ordered the helmsman to change course. All the masts on the bridge were lowered except the periscopes. Blowers were turned on. A sailor was sent up the ladder to the bridge with a video camera to film the intruder. In a real encounter, the pictures would be sent back to Naval Intelligence or used as evidence if the trawler collided with the sub, which occasionally happened during the Cold War.

Kinman began whispering excitedly into his pen mike to Brown, who was on the other side of the room standing in front of the radar screen. Kinman's eyes darted around the room, taking in the reaction of the crew to the approaching AGI ship. Quickly, he scribbled notes onto his steno pad. He was pumped up again. Of course, he was always pumped up. Hell, they were about to go out on patrol and play the world's greatest game of hide-and-seek. Who wouldn't be excited?

Kinman had become somewhat of a *Nebraska* legend in the two and a half years he had served on the sub. Stories about him had even spread to higher-ups in the squadron. Young Trident submarine officers were by and large a bland lot—quiet, introverted, practically all of them cerebral engineers. Kinman was not. He was loud and boisterous, impulsive and headstrong, the hard charger and bolt of lightning. He was the center of attention among the officers, an assault on their senses, the glue who kept the wardroom together emotionally. Built like a small lineman, he had a barrel chest and thick arms, and his 220 pounds strained the seams of his poopie suit. He was a twenty-seven-year-old bull in the china shop.

The sailors who served under him loved him because he was compassionate and fair with them. And Volonino thought he was a very capable officer, fun to lead. The captain did think that Kinman was a "management challenge" at times; it was a job to keep all that energy channeled down one beam. "Brent," Volonino liked to tell him, half joking, half exasperated, "you always have a full bell on"—an expression describing a submarine at top speed on the surface—"but no control of your rudder."

Kinman had never known his real father, a war hero who had earned the Bronze Star and Silver Star but died in a helicopter crash in Vietnam when Brent was just two. Sheryl Kinman eventually got remarried, to a quiet professor at the University of Georgia, but young Brent always had the explosive enthusiasm of his biological father. Every now and then, tears welled up in Sheryl's eyes when Brent did something outlandish or acted stubborn. The mirror image of his father, she thought.

Kinman was mostly an A student in school but was constantly called down by teachers for talking too loud in class. He won four varsity letters in high school, but hated theoretical classes. He found learning the airy concepts of calculus to be excruciatingly difficult, which probably explains why the toughest thing he had ever done in his life was becoming a submariner. He entered the Naval Academy and found it easy. What Kinman lacked in brainpower he made up for with hard work, which Annapolis valued. But never had he struggled so much as in mastering the technology of a nuclear submarine.

After graduating from the academy in 1994, Kinman followed all aspiring submariners and packed himself off to the Navy's nuclear power school in Orlando, Florida. He thought it was the hardest school on earth, period. Kinman had majored in oceanography at Annapolis. Most aspiring submariners majored in engineering, and that gave them a leg up. But Kinman was behind the first day he set foot in the nuclear power classroom. It was mental boot camp. For twenty-six horrid weeks, instructors stuffed math, nuclear physics, reactor plant systems, fluid dynamics and thermodynamics, electrical engineering, and radiological chemistry into his head. It all went so fast he thought he would explode. Much of it was theory, always difficult for him to comprehend. He studied fifty hours every week on top of the forty he spent sitting in classes.

Instead of being in the top 20 percent of his class, as he had been at the academy, Kinman struggled to stay at the lower middle in nuclear power school. Everyone around him seemed to be a whiz kid, a meticulous planner, a deep thinker, conservative, reserved, studious, inquisitive, bespectacled. Not Kinman. Except for the glasses, which he began wearing because he had to read so much.

Next came another twenty-six weeks at nuclear prototype school in Charleston, South Carolina, where students actually operated a nuclear reactor to master its mechanics. This was infinitely easier, because Kinman was learning by putting his hands on a piece of equipment, pulling its knobs, turning its switches. Then, after a year of studying nuclear energy, Kinman reported to submarine school in Groton, Connecticut, for three months to learn how to dive the ship, operate the periscope, fire torpedoes and missiles, maneuver in battle.

That was the best three months of his life, playing the hunter and getting a taste of the espionage side, which until then he had only read about in Tom Clancy novels. Up to that point, submarine training had been an intellectual exercise full of numbers, theorems, and rules. Now he was learning the art of war, which he found far more interesting. That was the problem with today's submariners, Kinman thought. They were technocrats more than warriors. The average lieutenant riding these boats considered himself a nuclear engineer first and a submarine officer second. "It almost feels like we're out there just driving the reactor around," he told himself. Too much thinking, not enough feeling.

Kinman found this ironic. He was sailing on a "boomer" whose routine hadn't changed one bit since the end of the Cold War. Yet the enemy had largely faded away, he knew. He had always wanted to be in the thick of battle, like the last generation who played cat and mouse with Soviet subs, or the generation before that who attacked convoys and dueled with German U-boats. He would never have that taste of adventure, he realized sadly. Not, at least, in the world in which he now lived.

Freeland began flipping through a white binder stamped CONFIDENTIAL and containing intelligence on Russian AGI ships to see if he could match the vessel that now confronted him with information the Navy had collected on surveillance crafts. There wasn't a real AGI ship out there. For the drill, Shawn Brown had pulled out a photo of one and shown it to Freeland. Dan Montgomery, the sub's communications chief, had also walked in with a copy of *Jane's All the World's Ships*, a civilian publication listing practically every vessel in the world. He began paging through it with Freeland to try to identify their mystery craft.

Brown, meanwhile, began whispering bearing and range numbers into the ear of Greg Migliore, a radioman who for the moment was operating the control room's radar console. For training purposes, Brown wanted Migliore to announce that he had spotted the vessel at another location on his screen. The new coordinates were generated for Brown from a laptop computer program that simulated the AGI ship on the prowl. Migliore announced that the ship was still closing on the sub.

Kinman knew that Brown was the brains behind the drills they were running. Kinman's job was mainly to make sure the scenarios Brown thought up didn't endanger the ship. Otherwise, he let the chief do as he pleased. Brown, who was as big as Kinman, certainly deserved the leeway, considering that at age thirty-eight he was the chief who had been aboard the *Nebraska* the longest. This would mark his twentieth patrol on ballistic missile subs. Brown estimated that he had spent at least five years of his life underwater.

The son of a truck driver who drifted from job to job in the West, Shawn had never spent more than a year in one high school. When he enlisted in the Navy in 1977, he ended up in submarines by accident—he had signed his name on the wrong form.

It was the best mistake he'd ever made. It had been a little scary for him at first, climbing into a confined space, never knowing for sure if everything would work after they closed the hatch and the sub dove. But he got used to it and quickly rose through the ranks as a submarine electronics technician, becoming a chief thirteen years later.

In the Navy enlisted rank structure, first there are three levels as seaman, then a sailor serves as a petty officer third class, second class, and finally first class, the highest rank before becoming a chief. But the

move from petty officer first class to chief is a giant
leap. A sailor who is smart, works hard, and keeps his
nose clean can reach the rank of petty officer first class
fairly easily. But often that is when his Navy career
grinds to a halt. Only a small percentage of petty officers first class ever make it to chief, and, as the Navy
has reduced its numbers, even fewer are likely to be
picked in the future. Sailors complain that it is easier to
become an officer than it is to move up the next enlisted rank to chief.

Chiefs are an institution peculiar to the Navy. In the
Army and Air Force, enlisted men who move to the
equivalent rank simply add another stripe to the sleeves
of their uniforms. Navy chiefs wear an entirely different uniform—khakis, like the officers, instead of the
dungarees or sailor suits of the lower enlisted ranks.
Sailors gripe that they become prima donnas with the
new uniforms and turn their noses up at the enlisted.
All Brown knew was that when he went to his supply
office as a petty officer first class in 1989, the chief who
was running it blew him off when he asked for help.
When Brown went to the office the next year wearing
those khakis as a chief, the supply chief jumped out of
his seat and couldn't do enough to help him.

As a chief, he became part of an exclusive club in
the Navy, a close-knit fraternity even with its own initiation rites. "The chiefs run the Navy" is a phrase
heard often in the service, and never is it more true
than in a nuclear submarine. Totaled up, the sixteen
chiefs on the *Nebraska* had centuries of experience operating its complex equipment. Officers coming aboard
wouldn't reach that knowledge level until they were
senior lieutenants or lieutenant commanders. Until
then, they gave orders and were addressed as "sir" by

the chiefs. But no one had any doubt who knew the most about the boat. Chiefs have their own way of dealing with a smart-ass officer who thinks he knows everything; they just stop giving him advice, and soon he will make a mistake that lands him in trouble with the skipper.

After a Navy board finally selected him to become a chief, Brown then had to go through a three-month "initiation" by his peers. He was given a "charge book," with intentionally confusing instructions on how it should be filled out, and then told to bring it to eighty chiefs on the base for signing. He had impossible-to-remember passages he had to recite before each chief. "I'm chief selectee Brown. Even though I'm not worthy to be in your shadow . . ." It went on and on. He did thousands of push-ups and ran for miles and miles. The initiation got him in shape and forced him to meet other chiefs he hadn't paid attention to as a petty officer.

Finally, after a week of formal classes on the personnel duties he would now have managing sailors, Brown began Initiation Day on September 16, 1990. He would never forget it. He was scared by then. The U.S. Navy had selected him for chief, but was he really ready to join this honored caste whose traditions had been established over centuries? Could he measure up to what was expected of him now? He had made so many mistakes in filling out his charge book that he wondered whether he was. The rites of Initiation Day were kept secret. Officers were not invited. But most of it was spent with harmless fraternity-like harassment: chiefs screaming in his face, push-ups, crawling in the mud, chugging alcohol, and sucking down raw eggs.

Exhausted, he finally stood at attention in a room

wearing his new khaki uniform before a court of veteran chiefs. Other chiefs stood in a circle around him.

One of the chiefs from the court stood up and began reading from a sheet of paper. "During the course of this day you have been caused to suffer indignities, to experience humiliations," he began solemnly. They may have seemed pointless, but "there was a valid, time-honored reason behind every single deed, behind each pointed barb." The responsibilities of a chief were unique in the Navy "and only in the United States Navy. . . . Your entire way of life has been changed. More will be expected of you, more will be demanded of you." This was why in the Navy, unlike the military's other branches, we "maintain with pride our feelings of superiority once we have attained chief petty officer."

The chief walked around the table and shook Brown's hand. Some of the old veterans standing in the circle had tears running down their cheeks. Brown stood there too dazed to comprehend the words of the creed being read to him. Only the next time, when he stood in the circle with the other chiefs to watch a new set of inductees receive the last rites on Initiation Day, would tears come down his face.

Freeland and Montgomery, the communications chief, finally identified the photo Brown had shown them to simulate the AGI ship that was on their tail. It was a Russian Vishnaya 864-class intelligence vessel. Hardee, who controlled the sub's movements from the bridge, decided to outrun the vessel.

It was after seven o'clock in the evening when the surface drills finally ended. Sailors had begun dismantling the equipment on the bridge and lugging it back down the ladder to be stored inside the ship. Evening on the first day out, particularly after dinner, was always the gloomiest for the crew. Meals at home with their families were still fresh in their minds.

The ship's junior sailors were hefting most of the heavy loads on their backs down the ladder. Some were fresh from the Navy's technical schools, while others had been sent directly from the five-week basic enlisted course at submarine school. Part of their time was spent acquainting themselves with the *Nebraska*'s machinery. But much of a new sailor's tour was consumed by the sub's menial chores, like washing dishes in the scullery. They called that "cranking."

Scott Shafer cradled a pile of metal poles under his arm and carefully lifted them down the control room steps to the second level. Though one of the sub's newest seamen, he was no kid. Actually he seemed out of place. He looked older than his twenty-four years. His head seemed too large for his body. He had wide eyes, thick lips, and a receding hairline. He spoke with a soft Texas drawl. It was a calm, serene voice that told you he had seen much of life already. In fact, Shafer was starting life all over again. The other rookies quickly realized the sub wasn't a cruise ship; they had already begun to gripe about the crappy jobs. Shafer couldn't have been happier. All work, he deeply felt, was honorable. Even cranking.

Life had been hard in Longview, Texas, a hot, dusty town near the Louisiana border. His father was a welder, his mother a nurse. They divorced when he was twelve, and though he lived with his father, Shafer was left to practically raise himself. By age seventeen, he had moved out of his father's house and settled into an apartment on his own, even though he hadn't finished high school. The year before he had met Sandy Gertz in math class. She had also come from a troubled home. Sandy now decided to move in with Scott. They had fallen madly in love, drawn together as well because both their families seemed to them to be so unstable.

Scott's high school principal and teachers were understanding. They gave him leeway on his class schedule so he could also work full-time at a plastics plant to pay his bills. But Scott and Sandy were still two teenagers thrust into adulthood, and the pressures began to build. Scott finally finished high school at age twenty and enrolled in a nearby college; both he and Sandy worked jobs to pay for his tuition. Shafer's dream was to be a law officer. But three and a half years after they had been living together, Sandy was pregnant. Blake was born on November 30, 1994. A year later, Scott and Sandy separated. They could no longer survive the strains of living together. The bills, college studies, raising Blake, working another job—it all became too much.

Scott quit college. He shared caring for Blake with Sandy and hired on with a security company, thinking that might be his doorway into law enforcement. It turned out to be a dead end. For three years, Shafer struggled as a security guard, changing companies four times because each went bankrupt. He barely made enough to take care of Blake. At one point, he lived in his pickup truck to save rent money.

Finally, in March 1998, Shafer enlisted in the Navy. It was the only way he could see to make enough money, to receive decent medical benefits, and to have some financial security in his life.

The Navy was heaven for Scott Shafer. Three square meals a day, a bed to sleep on, a roof over his head, and a paycheck of more than $1,100 a month, a part of which he mailed to Sandy in Texas each month for Blake. (Sandy by then had married another man, whom Shafer liked.) Shafer found Navy boot camp fun and the work in machinist school afterward not nearly as hard as slaving in the civilian world to make ends meet. The teenagers around him were always terrified of the instructors. Shafer would chuckle. God and the Navy would never put anything on us that we couldn't handle, he believed. The younger sailors didn't appreciate what the service was giving them in return for their hard work. He had no interest in going with them to bars on liberty. His focus now was on becoming a good sailor and raising his four-year-old son.

The *Nebraska* was such an amazing machine, Shafer marveled when he came aboard. It was as close as he thought he'd ever get to a space vehicle without leaving the ground. Making his way through its machinery rooms was like crawling through an iron jungle. He was in awe of the technology behind all those control panels, technology that provided unlimited energy from the reactor and unlimited destructive power from the nuclear warheads. The submarine was so huge, yet so finely precisioned. He would stop and run his hands admiringly over the smooth, sharp edges of the thick steel hatches that kept tons of water from pouring into compartments.

Even the crew radiated a sense of energy that he could feel the minute he climbed down the topside hatch. He pictured petty officers and chiefs yelling at him his first day aboard. Don't touch this or you'll blow up the ship! He was assigned to the machinery division, the "A-gangers" as they were usually called, which he knew could be a rowdy bunch. But no, they just handed him a rag and told him to start cleaning. The A-gangers were undermanned.

He was surprised. The petty officers threw less work at him than what he had been given in boot camp. They seemed far more laid-back. It shocked them when he'd arrive on watch an hour early to get a head start. As fast as he could find petty officers to teach him, he began mastering subjects from the submarine qualification booklet they had given him the first day. People would pay thousands of dollars for the education he was receiving on the *Nebraska* for free, so Shafer vowed not to waste a minute in soaking it up.

When he finally ran out of energy, usually in the early-morning hours, he collapsed onto a bed no better than for a stowaway. The petty officers and senior sailors slept in racks wedged between the sub's twenty-four giant missile tubes on the third level. They were hardly luxurious accommodations, but far better than "the ghetto," which was what the crew called the beds on the second level of the missile compartment above, where the junior seamen slept. Up there, Shafer's bed lay in front of tube twelve, tucked between two four-foot-high metal lockers with warnings on them that they stored hazardous materials. Two chest-high oxygen bottles stood at the foot of the bed. He could lift up the mattress and store his toiletries and spare uniforms in the "coffin locker" underneath it. A Naugahyde sheet was draped over the bed's top, because part of it stuck out into the passageway and sailors walked over it. With a curtain pulled across the foot of the bed, however, Shafer thought it made a cozy bedroom. The guards walking the passageway constantly to check the missile tubes were careful not to kick the beds and wake him up.

Never had Shafer felt so alive, so free. He kept so busy, he had smoked only two cigarettes the day before, when he normally would have gone through a pack. Maybe they'd already juiced the air with more oxygen. He would finish a work shift and want to stay for another eight hours. Even the grunge chores like packing garbage to be shot out of the ship didn't bother him. Every job was important, and Shafer was convinced he had begun making a difference the minute he climbed into the boat. This was what he wanted to do for twenty years, sail in subs for as long as they would let him. Never had he felt so right about anything in his life. Everything was finally falling into place.

4 • "Dive, Dive"

Shafer had finished lugging the metal poles down the control room's ladder to the lower levels. The sailors who had been on the bridge disassembling its phone sets and electronic receivers had all climbed down the ladder and shut the upper and lower hatches in the sail that kept water out. It was a few minutes before eight o'clock in the evening. The sun had set and the *Nebraska* was in deep enough ocean east of the Atlantic Coast that it could now safely submerge. Ryan Hardee was now the officer of the deck in the control room.

The son of a family doctor, Hardee had entered the Naval Academy to escape Dimmet, Texas. Everyone he'd known up to that point lived and died in Dimmet. He didn't want to join them. The Naval Academy was all the regimented misery that people had warned—"a great place to be from," as its alumni would say, but a grind for four years. The only thing that had made it bearable for Hardee had been Lori, the girl with stunning auburn hair and sky-blue eyes whom he met his junior year and who later became his wife. After dark, Lori would sit in a car on the other side of the wall that surrounded the academy and wait for Hardee to scale it so they could sneak away for the night. Hardee's roommates would stuff pillows under the blanket of his bed so that during rounds it appeared that he was still there.

Hardee looked like a younger version of Volonino: the same build, the same cut of hair. In fact, he had served under Volonino before, when he was a midshipman on a summer cruise and Volonino was the executive officer of the USS *Phoenix,* a fast-attack nuclear sub. Sailors and officers often alternated duty between the Navy's fast-attack and Trident subs. The fast-attack boats fired torpedoes at enemy ships and Tomahawk cruise missiles that could hit targets on land, and were just as stealthy but speedier than the Tridents. The fast-attacks were far more cramped inside, however, with sailors often having to "hot bunk," which meant two slept in the same rack on different shifts. The fast-attack subs also had far more erratic schedules, being dispatched at a moment's notice to distant spots for long periods whenever there was a crisis. The Tridents—the fast-attack sailors called their crews "boomer fags"—were underwater cities by comparison, with patrol schedules as predictable as sunrise and sunset. The *Phoenix* crew had treated the midshipmen like royalty, and Hardee remembered Volonino as its biggest rah-rah, always high-energy around them, about the same way he was now.

Two days after his honeymoon with Lori in 1996, Hardee had climbed aboard the *Nebraska,* at first so lost on the sub he didn't know which way was forward and which was aft. Kinman, who was already aboard and senior in rank, immediately pulled him aside and told him the tradition aboard the sub was that as the bull ensign, he had to shave his head when they set sail. Gullible, Hardee went bald, only to find out from cackling officers who ran into him later that Kinman had just invented the ritual. Fine, Hardee decided. He'd make sure every future bull suffered the same indignity.

Hardee reached for the 1-MC microphone hooked above the conn and announced on the speaker system that could broadcast messages throughout the ship that the submarine would soon submerge into the ocean.

For two hours, the crew had been rigging the *Nebraska* for the dive. That had set in motion a complex ballet throughout the boat. Submerging a submarine always has the potential for being a dangerous operation, particularly after sitting in port for a lengthy maintenance period as the *Nebraska* had. Equipment is creaky from not being used. Seawater leaks can be a problem, because seals have dried and shrunk and valves have not been under pressure from the ocean for a long time. And like an airplane that is landing, a submarine is at its most unstable when it dives underwater. Tens of thousands of valves, switches, and control knobs have to be in the correct position for the dive. A malfunction or improper setting with any one of them and the crew could lose control of the sub during the descent or, even worse, let in the enemy—water. There is no way to inspect all of the items before the dive, so Volonino, like all sub captains, designated crewmen to check at least several hundred that were absolutely essential for the sub to sink safely into the ocean.

An officer and petty officer were assigned to each compartment in the sub, with lists of equipment to check then double-check before the dive. When they finished, the two men signed a "rig for dive status book" in the control room that certified their space was ready. Volonino then audited the book before a dive to satisfy himself that his boat would submerge safely.

Volonino walked briskly into the control room shortly after eight. He stopped at the primary plot table to check the sub's location off the coast, then flipped open the dive status book at the conn. Also, it had all the signatures from the two-man teams in each compartment. His sub was ready to go under.

The lights in the control room were dimmed so that Hardee's eyes would not have to adjust to the dark outside when he peered into the periscope. Also, if the control room wasn't darkened, the light from it could be reflected up the periscope's mirrors so the part of the scope sticking out of the water would become a flashlight giving away the sub's position to surveillance planes. The control center now was lit only by the red, green, yellow, and blue glow from indicators and screens on the panels.

Volonino could feel every eye in the control room staring at him as he finished leafing through the status book's pages. It had taken him some getting used to at first. Always being the center of attention. Always the one everyone on the ship looked to for answers, for decisions. It could be heady, yet intimidating and lonely.

Nobody feels ready to take command of a modern nuclear sub. Volonino believed that anybody who did was an idiot. He had been put on the career fast track for command—ironic, he realized, considering that after his first four years in the service he had toyed with the idea of getting out and taking a civilian job. But he had remained, captivated by these steel tubes and the sea, and the Navy had rewarded him with coveted positions as a sub's engineer, then as an executive officer. Along the way there had also been choice staff jobs in Washington, plus a stint at the Defense Department's prestigious War College. Before taking charge of the

Nebraska on April 11, 1997, he had spent fourteen months in commanders' school to prepare for the position. But it didn't relieve the unease he felt the first time he walked across the gangplank and onto *his* sub.

To Volonino's tremendous relief, his first week aboard the *Nebraska* had been far less overwhelming than he had feared. Other skippers had warned him of the phenomenon, but you had to experience it for yourself to really appreciate it, he now realized. The minute he walked aboard, he mysteriously had instant credibility with the crew. He hadn't earned it. The more than 160 sailors and officers on the sub didn't know him from Adam. But not a single man viewed him as a fledgling commander, a rookie they'd have to suffer until he learned the ropes. Volonino, to his surprise, was now the captain of the ship, and every crew member immediately gave him his loyalty, 100 percent.

The crew considered him just as much their property as he considered them his. Sailors, he found, had high standards for the captain of their ship. They wanted him to be not only technically competent but someone they could look up to. They wanted their captain—in fact, they seemed desperate for their captain—to be the best skipper on the waterfront. They collected information on him to support that view, and they were surprisingly gracious in overlooking his early deficiencies. It made it difficult for him to do anything profoundly stupid in those first few weeks. A crew could eventually turn on a skipper if he became a tyrant. But Volonino was struck by how much automatic respect a commander got in the beginning.

You had to be careful how you acted with that instant credibility, he quickly found. Be careful what you ask for, because you might get it. In fact, Volonino

soon learned that you didn't even have to ask and the
crew would start jumping through hoops. If he walked
down a passageway and glanced at something out of
place on a piece of equipment, sure enough, someone
nearby would notice that the item had caught his attention. When he returned a half hour later, he'd find a
half-dozen sailors taking the equipment apart and
scrubbing it. The chiefs amplified down the chain of
command every remark, every gesture he made. In his
first couple of months, he didn't realize how many
lightning bolts he shot out. He soon learned he had to
watch what he said or did so people wouldn't overre-
act. Above all, he had to be extremely gentle with
sailors. Criticize them too harshly and they would
practically collapse on you.

Hardee made one last check with the navigation
sailors, who took a fathometer sounding that showed
the ocean bottom was far enough down that the sub
had no chance of hitting it during the dive.

"Captain, request permission to submerge the ship,"
he asked. Hardee reported the position of other vessels
near them and the condition of the sub for the dive.

Volonino pressed his eyes to the periscope and
swung it around full circle for his own look outside.

"Very well, officer of the deck," Volonino said after
looking through scope. "Submerge the ship."

"Submerge the ship, aye," Hardee repeated, following his part of the script. He turned to the diving officer
standing behind the two sailors who would steer the
sub. "Dive, submerge the ship," Hardee ordered.

The diving officer aboard a Trident was always a
seasoned chief, in this case Jeff Spooner from the
weapons department, who was also the sub's computer
guru. He controlled the two sailors who steered the

sub, plus the chief of the watch to his left, who could pump water on and off the sub's huge ballast tanks to make it heavy or light.

There was a ritual to how commands were issued in the control room. Though everyone was within spitting distance of everyone else, orders were passed down the chain of command in a strict manner, and no one jumped the line to talk directly to the person carrying out the task. Spooner now turned to Stacey Hines, another chief in the weapons department, who was serving as chief of the watch and sitting before the ship's control station.

Hines had a big job. The 127 dials, knobs, and switches on the station's broad panels, which he had to know how to operate blindfolded, constituted the main hub for the submarine's mechanical operations. Not only did he control the ballast system from it, Hines monitored the hydraulics plant, operated the ship's hovering system, sounded emergency alarms if the vessel got in trouble, raised and lowered electronic masts and antennas, spooled out the towed sonar arrays, serviced the air banks, manipulated the trim system, tracked the status of missile firings, controlled the internal communications system, and, if he had to, surfaced the ship quickly in an emergency by pulling back two heavy metal levers over the panels nicknamed the "chicken switches."

Hines also ran the ship's messenger service and the wake-up calls for the crew. Very little happened on the submarine that wasn't cleared through the chief of the watch.

Spooner ordered Hines to sound the diving alarm and make the announcement on the sub's speaker system that the ship was about to submerge.

Hines reached for his microphone and announced quietly: "Dive, dive." Then he twisted the brass diving alarm lever in front of him to the left. It let out an "*ooga, ooga*" that sounded like a sick pig squealing throughout the ship. He reached up to two switches on the ballast control panel to open the sub's main ballast tanks' forward and aft vent valves so air would be blown out and tons of water would be sucked in. That would make the sub heavier, so it would sink gracefully into the ocean. Hines brought the microphone to his lips again and announced into it, "Dive, dive," once more.

But something was wrong. In the top right corner of the ballast control panel, a thin blue horizontal light on the vent indicator for the aft group of ballast tanks remained lit. And the orange circle light above it failed to illuminate. That meant the vent valves for giant ballast tanks in the rear of the sub hadn't opened to let in water. The vent indicator lights for the forward group of ballast tanks illuminated as they were supposed to on the panel, which meant the giant cavities in the front of the sub were filling with water. But having the front of the boat full of water and the back full of air was the last thing Hines wanted. The sub would start to nose-dive dangerously into the water and the crew could lose control of the ship.

Almost in the same instant that he spotted the problem with the vent indicator light, Hines shouted, "Aft group failed to open. Recommend securing the dive."

"Secure the dive," Hardee quickly ordered. Hines turned the vent switches to stop the forward ballast tanks from taking in water.

"Officer of the deck, we're at four-one feet and holding," Spooner announced.

"Report the reason the dive was secured," Volonino ordered. But he already knew. Kinman and Brown were standing at far corners of the control room with their red drill team caps on whispering into pen mikes. They had intentionally caused a fault in the main ballast tank vent-control system to test how the crew would react. This was another drill. The *Nebraska* had six large ballast tanks that encased the hull, with flood grates underneath the sub to let water rush in. Three of the tanks were in the front of the sub and three were in the back. When the boat was surfaced, the ballast tanks were filled with air; the vessel, in effect, was surrounded by a giant bubble that kept it buoyant. When the sub submerged, vents at the top of the ballast tanks—each tank had two vents—were opened, which forced the air out of them so water could pour in from the bottom grates. Brown had had his sailors on the drill team rig the circuit for the six rear vent valves so Hines couldn't open them electronically by flipping the switch on his control panel.

A-ganger sailors on duty for the dive rushed back to the engine room, where the control mechanisms were located. The minute he had heard the problem announced on the sub's speaker system, Chad Thorson, who was already posted in the back of the ship in the engine room, began troubleshooting. Within minutes, Thorson thought he had spotted the problem. The electronic circuitry for opening the vent valves was screwed up.

Harry Ganteaume, the sub's engineer, sprinted up the ladder into the control center. Thorson wasn't far behind him, clutching a six-inch-thick manual with instructions on how to repair the circuit. As the ship's engineer, fixing the stuck valves was Ganteaume's re-

sponsibility, and when Thorson radioed him where he
thought the problem lay, he had a solution in his head
by the time he reached the control room.

Ganteaume huddled with Volonino on the conn and
spoke in a soft, calm tone. "Manual override" was all
the others in control heard from the three officers
bunched together. It would take a half hour to fix the
circuit, Ganteaume told Volonino.

In the meantime, however, the skipper wanted to
submerge his boat. To do so would require a manual
override. There were several ways the crew could mechanically open the vents instead of having the electronic switches do it. Ganteaume, with Thorson
hurriedly paging through the manual to recite the procedures, explained to Volonino how he would manually override the hydraulic control valves.

Harry Ganteaume was the quietest workaholic his
shipmates had ever met. He hustled about the sub in silence, entered the control center, the wardroom, every
meeting he attended, always unobtrusively. He just
seemed to be there, the kind of person who blended
with the background. Slender, with deep brown eyes, a
pointed nose, and thin black hair receding from his
thin forehead, Harry had a gentle manner to him. When
he spoke, his words, wrapped in a rich Spanish accent,
conveyed a serenity. He was unfailingly kind and def-
erential, modest to a fault. He had a warm smile, but
only occasionally did it interrupt the tired and worried
look on his face from the crushing workload of managing the biggest and most demanding department on the
submarine.

Senior officers who had come aboard for visits had
pulled Volonino aside and asked whether his engineer

was weak because he was so unassuming. Volonino assured them that he most certainly wasn't. In fact, Volonino told them that Harry was one of the best engineering officers he had ever met. He called him his "quiet giant." The some sixty men in his department revered Ganteaume—he had an encyclopedic knowledge of the ship's engine room, and he treated them fairly—although they considered him the sub's mystery man because he was so quiet and they knew little about his personal life.

Volonino had practically ceded to him the running of the back of the ship because he trusted him so. He had been "spot promoted" to lieutenant commander ahead of his contemporaries—which always occurred in the engineering post, because it was such a big job—and at the rate Harry was advancing, Volonino felt sure he would be the captain of his own Trident one day. Which would be a historic achievement of sorts. Thirty-six-year-old Harry Ganteaume would be the first Venezuelan native to command the most destructive war machine in the world.

He had been born in Maracay, a medium-sized city an hour's drive east of Caracas, the Venezuelan capital. His father, also named Harry, had come from a tobacco family and had worked all his life for Philip Morris. In 1979, the family had moved to Richmond, Virginia, Philip Morris's corporate headquarters, where Harry Senior worked in the international division for Latin America. Young Harry, a shy fifteen-year-old by then, arrived knowing little English and, with his three sisters, was enrolled in private school. Within six months he was functional in his new language, albeit with a heavy accent. But it took longer to acquire new friends.

Richmond was a closed society, whose old families and old money shunned outsiders. It took more than a year for him to make friends in school.

During the next two decades in Richmond, Harry Senior never gave up his Venezuelan citizenship. Though Venezuela had become economically unstable, he always dreamed of returning one day and living his retirement years in his homeland. His wife, María, who had no desire to return, became an American citizen along with the children. Young Harry, who by then no longer considered himself a Venezuelan, swore his oath of allegiance to his adopted country in 1986, his senior year at Bucknell University in Pennsylvania. Ganteaume, who majored in mechanical engineering, chose Bucknell because it was a small school where he thought he wouldn't be just another face among the thousands. It was for the same reason that he chose to be a part of a small submarine family rather than a number aboard a big impersonal ship.

Ganteaume took well to the Navy's rigorous nuclear power training. Math and science came easy to him, and he had always been mechanically adept. But in 1995, when he had reached the rank of lieutenant and was a shift engineer at the Navy's nuclear prototype school in Charleston, Harry resigned his commission. His commanding officer at Charleston was a despot who ruled by intimidation and delighted in dressing down subordinates in front of others. The CO got results, but his men performed out of fear.

Ganteaume could have suffered the tyrant and moved on to his next duty station. But then what? he asked himself. He was not a screamer. He could count on one hand the number of times he had ever raised his voice. And never would he consider personally offending a subordinate no matter how mad he became. But if being a bully was what it took to be a commander in the United States Navy, Ganteaume wanted no part of the service. He went to work for the General Electric Company as a maintenance engineer in Albany, New York.

He returned to the Navy six months later, however. The GE executives in Albany were friendly, and they went out of their way to make him feel at home. But it was a boring nine-to-five job. There was nothing unique or special about what he did. He was an engineer in a division that manufactured plastics for dashboards, not quite as glamorous as sailing the seven seas in a nuclear sub. The quality standards in the plant were not as rigorous as with military reactors, and he missed the closeness of his submarine family.

There was another reason, just as important, for why he wanted back in the Navy. In the months before he had resigned his commission, his boss from hell had been replaced by an officer who was a firm leader but a gentleman. Ganteaume had begun to realize that you could in fact command sailors in this service but still treat them with kindness and respect.

The Navy was eager to have Ganteaume back. The service was always short of young submarine lieutenants, because so many left the service at the end of their first hitch. Now Harry wanted to become a submarine captain more than anything else. He had arrived at the *Nebraska* in the fall of 1996 to take over the job of ship's engineer. It was considered the third-highest position among the officers on the boat, behind the captain and the XO, and more prestigious than heading up the navigation or weapons department. A sub's engineer usually had a leg up in screening for the two biggest prizes: executive officer and then commander of your own ship.

Harry devoted practically his entire life to the job. During the overhaul period before the sub set sail he worked eighteen-hour days for weeks overseeing equipment repair and training his men for the patrol. (He tried to reserve Sundays for church and a round of golf.) He had never married and had decided to set aside having much of a social life during the hectic early years of his career. Maybe there would be time for a woman in his life later. But not now. This was his last patrol with the *Nebraska*. He would leave the sub for a year of graduate school at the Naval War College in Newport, Rhode Island. Then, after a staff job, if he was lucky, he would be selected to be a sub's executive officer, the next critical step to becoming a captain. His father, who had retired in Richmond and whom young Harry worshiped, read every book he could find on submarines after his son joined the service. He would have been so proud of him.

Would have been. Three weeks before the *Nebraska* set sail, Harry's father died. A bolt-out-of-the-blue heart attack had felled him. It was a cruel twist of fate. Retired just two years earlier, he also had taken up golf and had begun enjoying life and the fruits of decades of labor. His death was devastating for Harry. Returning to the sub after the funeral, he plunged into its brutal work schedule hoping that would take his mind off losing his father. But the pain would always creep back during the rare moments he had to himself.

Fifteen minutes after the sub's dive had been suspended, Ganteaume's sailors were ready for the manual override of the control mechanism for the aft ballast tank vents. Hines opened them again. Keith

Larson, a young torpedoman who was the inboard helmsman and was operating the steering wheel to Spooner's right, reached for the engine order knob on the diving officer control panel to his left and twisted it clockwise to the two-thirds-ahead setting. That set off a bell, which signaled the engine room to increase the sub's speed. With his right hand, Larson kept his wheel centered so the rudder at the stern would remain in the middle as the boat submerged. Hardee, meanwhile, began peering through the number two periscope at the conn. He could see through the scope if the dozen ballast tank vents topside opened, because tall thin plumes of water and mist would gush out of them like geysers when the air was forced out. Hardee spun the periscope around so he could look toward the front of the sub. He saw the plumes.

"Venting forward," he announced to the others in control, his eyes still glued to the scope as he spun it around to peer at the back end of the sub.

"Venting aft," he announced. The manual override for the six aft vents had worked. Air was coming out of all the ballast tanks and water was pouring in.

Slowly the Trident began sinking into the ocean.

"Four-two feet, four-six feet," Spooner began calling out as the sub's hull sank deeper and deeper. He had a foot propped up on the bar in front of the diving officer control panel, an arm draped over his knee.

"Five-zero feet," Spooner continued.

"Deck's awash," Hardee announced, continuing to peer through the periscope. Water now lapped over the top of the hull.

Spooner tapped Jason Bush on the shoulder and ordered him to go to a "full dive" on his stern plane. Bush, a stocky, twenty-two-year-old Alabaman whose regular job was launching torpedoes from the control room's attack console, sat before the steering wheel to the left of Spooner, which controlled the planes at the back of the ship. Shifting those planes could point the sub up or down. He now pushed his steering wheel forward so the *Nebraska*'s nose dipped at about a four-degree angle. Spooner backseat-drove him, watching the "down bubble" light floating on the vertical gauge in front of him to make sure Bush kept the stern plane at the correct angle.

But as the sub's depth passed sixty feet, Spooner quickly noticed that not all the gauges were telling him the same depth. There were a half-dozen digital and analog depth gauges on the instrument panels in front of him. But the digital and analog readings now differed. Another gremlin Kinman and Brown had sneaked in to test whether the diving party would become confused on how far the sub had sunk.

Spooner wasn't fooled. "Officer of the deck, my analog gauges are not tracking with the digitals," he warned Hardee.

Spooner decided the analog gauges were still giving the correct readings and ordered Bush and Larson to watch them and ignore the digitals.

"Six-five feet," Spooner announced, then tapped Larson on the shoulder. "Full dive on the fairwater planes." Larson pushed his steering wheel forward firmly but slowly. Rookies who yanked the wheel back and forth too quickly were knocked for "slamming the planes." The fairwater planes of the sub's sail were now underwater and could join the stern planes in pointing the vessel downward.

"Seven-zero feet," Spooner said. Hines reached up to switches on the ballast control panel and shut the twelve ballast tank vents to stop any more air from escaping. He did that just in case there was an emergency at that point and the sub quickly had to surface. Otherwise, the air that might give the sub buoyancy in a crisis would just blow out the open vents.

"All vents shut," Hines announced.

The control room was silent and still dark as the sub continued to submerge. The only sound was Spooner announcing the depths they passed. The mood was always focused and tense in the control room when the sub submerged, even more so now because Kinman and Brown were introducing all these troubling faults as they dove. Volonino's eyes darted from gauges to digital readouts, on alert for any signs of problems with the crew's diving procedures. So far, there were none, so he kept quiet.

"Eight-zero feet," Spooner announced. The crewmen could feel the sub tilting down, so they leaned back slightly on their heels to compensate.

"Scope's under," Hardee announced a minute later. "Lowering number two periscope." He flipped up the black handles and the fat gray tube came sliding down. The *Nebraska* was now completely submerged.

"Nine-eight feet," Spooner announced. "One hundred feet."

With the periscope down, the lights were flipped back on in the control center. They all blinked their eyes. Murmured conversations began.

Spooner continued to call out the depths as Larson and Bush kept their wheels pushed forward.

"One-one-zero feet."

Hardee responded with "Very well, dive" after each depth call.

"One-two-zero feet. One-three-zero feet."

"Everyone stay focused," Hardee admonished. "We're still submerging the ship." The room became quiet again. They were still in what submariners considered no-man's-land, where the ballast tanks weren't completely full or empty and the vessel was in its most unstable state. Hines began whispering into his headphone set for reports from sailors scattered around the ship on whether they had spotted any leaks. Some had. The leaks were being simulated by the drill team, again to test the crew's reaction.

"One-four-zero feet," Spooner continued, whispering directions to Larson and Bush between depth calls.

Spooner ordered the two sailors to pull back their wheels. If the sub was going to level out at 160 feet, the stern and fairwater planes had to be flipped up well before that so the boat didn't overshoot and dive deeper.

"One-five-zero feet," Spooner announced.

Volonino silently mouthed the word "now" to Brown, who was standing on the other side of the control room. Brown covered his hand over his pen mike and began talking.

All kinds of alarms suddenly blared in the control room. Loud buzzers and bells. Practically all the illuminated dials at Hines's ship's control station blacked out. It seemed like a catastrophic power loss for the control room. All the electronic depth and course indicators were blank, hydraulic control valves powered from the station were dead. The normal electronic controls that moved the planes were out. For a brief second it was chaotic.

Spooner kept his cool and flipped emergency switches to keep control of the stern and fairwater planes so the sub could still be driven down. The mechanical analog gauges, which weren't powered by electricity, were still operating, so he ordered Larson and Bush to pay attention to them.

"Keep calm," he told them, his hands on the two shoulders. "We're fine with the analog gauges." A preacher's son, Spooner had spent twenty-two years in submarines. This was his seventh patrol on the *Nebraska* and his last. Spooner planned to retire and find a more lucrative civilian job in computers. He reached over and pushed both wheels forward for a moment to keep the sub descending so it would level off at 160 feet.

An electrician posted in the control room had rushed out to a nearby circuit box to hunt for what had caused the power loss. He quickly found it. Brown had tripped a circuit breaker, which shut off the power to the ship's control station. The electrician phoned the glitch to Hines and reset the breaker.

Dials and indicators again began sparkling on the station's control panels. Spooner switched back to regular power for the planes.

"Pay attention now," Spooner said to Larson, patting him on the back.

Larson and Bush eased their wheels back, and the *Nebraska* settled just short of 160 feet.

"Helm, right fifteen degrees rudder, steady course one-one-zero," Hardee snapped out the order from the conn.

Larson repeated the order, then turned his wheel to the right. When the sub reached a course of one hundred degrees, ten degrees short of the course Spooner had ordered, he spun the wheel to the left. That would keep him from overshooting the 110-degree course he wanted. A few minutes later, Larson centered the wheel so the sub lined up on 110 degrees.

Hines finally delivered his report to Hardee on the number of leaks the watch-standers around the ship had spotted. One of the leaks, in the missile section, hadn't been simulated by the drill team, however. Water was slowly seeping from the grease line going to the hatch over missile tube number one at about "two drops per minute," Hines announced. Nothing serious at this point. At the depth they were at, the water outside was pushing about four and a half tons' worth of pressure on them. There were always small, controlled leaks that had to be stopped the first time the sub dove after a long refit.

Spooner and Hines now busied themselves trimming the sub, a complex operation that involved pumping water through trim tanks scattered around the hull so the vessel would sail level through the ocean at neutral buoyancy. That was important for a number of reasons. A trimmed boat doesn't lean one way or the other, which the crew appreciates because it makes walking on the decks easier. Sailing level, the sub also travels faster and quieter through the water.

But many factors affect a sub's buoyancy and balance. Colder water is denser, making the sub lighter. The different layers of cold and warm water in the ocean can cause the sub to rise or fall. More salt in the water and the sub becomes lighter. The deeper the sub submerges, the more the hull compresses, creating less buoyancy, but the water also becomes denser, creating the counteracting effect of more buoyancy. How spare parts, food, and fuel are stored in the sub affect its balance underwater. The balance shifts as well during a patrol as the crew eats food and expels waste.

Spooner and Hines began draining and filling the

trim tanks until they got the *Nebraska* as horizontal as they thought they could get it.

"Officer of the deck, one-six-zero feet, trim satisfactory," Spooner finally reported to Hardee.

Because it has no keel, a sub on the surface is rocked from side to side more by the waves than a regular ship would be. Submariners hate sailing on the surface because many become seasick in choppy seas. But once submerged, a sub sails silky-smooth most of the time and the crew can hardly tell the boat is in water.

Calm returned to the control room. Shafer walked in carrying a metal stool with a padded back and installed it at the spot Spooner had been standing at so he could now sit as he monitored the two planesmen steering the boat. Spooner stretched his back and plopped into the chair. "That was a graduate-level drill," he muttered to himself.

It certainly had been. Volonino had ordered Kinman and Brown to "take the training wheels off" and throw particularly nasty problems at the crew when they submerged the ship. "If we suck, we suck," he had told them in the officers' study before the drill began, adding with a laugh: "Although I don't know how much more sucking I can take today." By now the entire Kings Bay base probably knew about the near disaster at buoy 41. The harbor pilot was sure to have blabbed to his buddies.

5 • The "Cob"

Ray Chesney could avoid it no longer. Nobody could explain to him—or at least not to his satisfaction—where this tradition had come from that the bull ensign had to shave his head. People just kept asking him when he planned to part with his locks. It was clear he had no choice, especially if he wanted to fit in and show he could be a good sport. After the boat submerged and the pace inside had slowed for the night, about a half-dozen sailors found Chesney and escorted him down to the crew's latrine in the missile compartment.

Shawn Olmstead, a petty officer second class from the electrical division, stood by a chair with clippers in one hand. Olmstead was one of the ship's barbers, but only because he had a set of clippers. And there was only one style he could cut—shaved. With a devilish grin, he patted the chair for Chesney to sit down.

The ensign slumped into the seat facing the metal sinks and mirrors. The sinks were all spotless, and they would stay that way throughout the patrol. Bathroom etiquette was strictly enforced on the sub, and visitors to the vessel were politely but firmly told the rules. Because there were so few bathrooms (the Navy calls them "heads") and so many people crammed inside, everyone cleaned after himself. A fat sponge sat by the sink to clean bowls after toothbrushing or washing. Squeegees hung in the shower stalls to wipe down water after each shower so the stall wouldn't become mildewy. A valve was attached to the shower head to turn the water on and off so it would be conserved. Proper technique: open the valve to let out water to wet down first, then close the valve and soap up, then open the valve again briefly to wash off the soap. If someone ever let the water run freely all the time, his shipmates called it a "Hollywood shower," which was frowned upon.

Olmstead draped a towel around Chesney and went to work as laughing sailors crowded around the bathroom's entranceway to watch the ritual.

Chesney had forgotten to bring any books with him for the patrol, which was just as well because he didn't know when he'd have a chance to read them. Ever since they both had come aboard the *Nebraska* during its refit, Chesney and Nowalk had been on the run nonstop. New sailors were given qualification cards, blue pocket-size cards with several pages that they had petty officers sign as they learned each part of the sub's operation. The "qual card" Chesney and Nowalk each carried around was much larger—a two-inch-thick notebook crammed with qualification sheets crewmen would sign. It would probably take them two patrols to become qualified submariners and to earn their gold dolphins. One sheet Chesney would have to get signed, for example, was for "rigging the forward compartment for dive under the instruction of a qualified officer." It could be obtained with several hours of study and about an hour of hands-on work. But even after a junior officer pinned on his dolphins, the qualifying wouldn't stop. He next would have to attend eight to nine weeks of school to qualify as a chief nuclear engineer officer. As he rose in rank, there would be more qual cards, and they would become smaller but far more difficult and time-consuming to get signed. For example, on Volonino's qual card to be the commander of a submarine, a single signature had been awarded when he could "demonstrate proficiency as an approach officer during twenty simulated torpedo approaches, maintain a record of each approach in a torpedo approach and attack log," conduct "at least four satisfactory approaches . . . in which actual torpedos are fired," and "complete an oral examination with the commanding officer on approach officer procedures, tactical considerations, limitations and measures for optimizing attack on multiple surface ship formations and submarines." It had taken him over two years to get that signature.

Chesney was also quickly discovering that the new ensigns on board suffered many indignities. They weren't given one of the officer staterooms in the forward compartment's second level. Instead, Chesney and Nowalk slept with other sailors in one of the nine-bunk compartments in the third-level missile spaces. On "Bull Night," Chesney had to fetch popcorn and pitchers of soda for the officers watching a movie in the wardroom. And the jokes had started the first day he arrived.

The officers were having a party at Freeland's house the night Chesney and his wife, Heather, drove into Kings Bay from Michigan. The first person Chesney saw at the door was Nowalk, who had arrived earlier and was in on the joke.

"There's been a sudden change of command,"

Nowalk whispered to him hurriedly. "They've got a new young captain and he's an asshole."

Steve Habermas, the sub's communications lieutenant, walked up to Chesney wearing a commander's ball cap with the scrambled eggs on its bill. "Let's go outside and talk," he said sternly.

Mystified that he hadn't heard about the *Nebraska*'s change of skippers, Chesney followed Habermas to the porch.

"What was your class rank at nuclear power school?" Habermas demanded. Chesney gave it, plus his rank for nuclear prototype school, which was slightly lower.

"Oh, so you're slipping," Habermas said accusingly. Chesney didn't think so.

"How much did you study in prototype?" Habermas continued like a prosecutor.

"A lot," Chesney answered.

"It must not have been enough," Habermas snapped, his voice growing louder as he moved closer to Chesney's face. "The *Nebraska* is the best boat on the waterfront. You're going to need to do better on my ship."

"I think I can," Chesney said, at a loss. Nowalk was right, he thought. This guy is a prick.

Kinman walked out to the porch, and Habermas angrily dressed him down for wearing a University of Georgia Bulldogs cap. Other officers came out and acted so obsequious that Chesney thought Captain Bligh had taken over the sub. Inside, Heather was beginning to fret because this new CO was obviously berating her husband out on the porch.

"Don't worry about Ray," one of the wives finally whispered into her ear. "They're just having some fun with him."

Habermas continued haranguing Chesney until finally he started to grow suspicious. No commander would be this outrageous at a party, or at least he didn't think so.

A smirk started creeping across Chesney's face. It happened when he became perplexed. He couldn't help it.

"You think this is funny!" Habermas said, almost shouting.

"No sir," Chesney said wiping the smile off his face.

Habermas couldn't keep up the act any longer. He broke out laughing, along with everyone else on the porch.

"Welcome to the *Nebraska*," he finally said, wiping his eyes.

Now, as Olmstead finished shaving Chesney's head, clumps of light brown hair lay around the chair. Chesney rubbed his hand over his stubbly bald head. The old smirk came over his face.

It wouldn't be so bad. The only people who would see him looking this goofy for almost the next three months would be the other crew members. Time enough, he hoped, for his hair to grow out. And as it was, he wouldn't stand out on the boat. Almost half the crew shaved their heads during the patrol. Even Nowalk joined him. The busy sailors preferred shaved heads because they didn't have to fuss with combing.

Dave Weller was having a bad hair day as well. The forty-year-old chief of the boat—COB—leaned back in the metal chair in the chiefs' quarters and propped his feet up on the lip of the couch in front of him. Standley was already there, snoozing. The chiefs called it a "Nauga nap." All the padded chairs and couches that lined the lounge were covered with light brown Naugahyde. The chiefs, who were up at all hours, would slip into the lounge at different times of the day to recharge with short naps.

It was something mysterious about those couches, the chiefs would say. Plop on them, close your eyes, and "the Nauga would suck all the life out of you." You were instantly asleep.

The chiefs' quarters, or the "goat locker" as it was sometimes unflatteringly called, was their private lair. No sailor entered the room, which was the size of a small anteroom, without asking permission. Volonino knocked before coming in (Boyd, the XO, had forgotten several times, and the chiefs had definitely noticed it). Padded couches were built into its bulkheads, where there were plaques, submarine photos, and a ship's clock. The lounge also had two card tables anchored into the deck, a coffee machine on a shelf in one corner, and two color TVs and a VCR in the opposite corner. Surrounding the lounge were the berthing spaces for the chiefs and their private bathroom and showers. It was their sanctuary, the one place on the sub where they could escape sailors and officers, where they could let down their hair and call each other by first names. A red sign on the door to the lounge warned:

What You Say In Here
What You Hear In Here
What You Do In Here
What You See In Here
"STAYS IN HERE"

Weller began griping to other chiefs who walked in and out of the quarters. Jason Dillon, a twenty-two-year-old storekeeper, had showed up the first day of the patrol with his hair cut Navy-regulation-short, but dyed a bright orange. With squadron riders aboard, Weller wasn't about to have a punk rocker strolling around the sub.

"I told him either to shave it off or dye it back to his natural color," Weller said. "And what does that fucker do? He goes to the captain to appeal it."

Volonino had an open-door policy for the crew, but orange hair wasn't something to bother him with. (Dillon didn't see why everyone was making a federal case about his. "It's not like anyone sees us out here," he had complained to his shipmates.)

Much of Weller's day was spent with this kind of personnel headache: ordering haircuts, overseeing the watch schedules, counseling sailors when they ran up debts, driving to the county jail in the middle of the night to bail them out. Actually the *Nebraska* had few discipline problems, far fewer than are found on surface ships. This is the case for most Navy submarines. A sub's sailors are all volunteers and among the best-educated personnel in the fleet, which means they are smart enough to stay out of trouble. Weller had no more than a couple cases each year of enlisted men running up debts. And only in the past seven months had the *Nebraska* gone through what for it was a rash of trouble: one desertion, three discharges for drug use or forging a qualification card, and about a half-dozen other men brought before a "captain's mast" for minor infractions, where Volonino meted out punishments like docking their pay.

No one had an explanation for the flare-up, except that sailors always get into more trouble in port, where they have more freedom and temptations. Once at sea, the problems almost disappear. A log might be filled out incorrectly, a seaman might talk back to a chief, but rarely anything more serious. There is little stealing during a patrol, mainly because with hardly any liberty ports to visit and so little space aboard the sub for personal items, sailors don't bring much cash or anything else worth stealing.

Yet even though the number of personnel problems on the *Nebraska* was small, Weller and the sub's officers agonized over them. Some sailors, in fact, griped that Volonino fretted too much and gave miscreants too many second chances instead of just booting them out of the Navy. But as a practical matter, the loss of even one sailor causes all kinds of administrative pains. There is no fat in a sub's roster (the *Nebraska* was already more than a dozen qualified sailors short for this patrol), so watch schedules have to be rearranged and crewmen have to double up on jobs when a sailor is kicked off. Replacements often don't come quickly. As a result, officers do all they can to salvage a seaman so they won't have to send him packing.

Discipline cases are also a barometer of crew morale, which Weller and Volonino watched closely. The weapons aboard were deadly dangerous, so they moved quickly to quell even the faintest signs of discontent. The sub would perform poorly if crew morale was low. On patrol, problem cases have to be resolved, because there is no place to hide the problem. The sub has no brig and few opportunities to helicopter off malcontents. When they returned from a grueling patrol, the officers and chiefs had to carefully calibrate work time and time off to keep frustrations vented. To raise

spirits, Volonino was constantly holding awards ceremonies to pass out certificates, plaques, and letters of
praise. To keep young sailors from quitting, the Navy
dangles cash in front of them. A nuclear power special-
ist, for example, can earn as much as a $45,000 bonus
if he reenlists.

Dave Weller had moved up the ranks quickly in the
Navy. He had become *Nebraska*'s chief of the boat, the
top post for any enlisted submariner, after just nineteen
years in the service. Weller had done well because he
realized early on that to get ahead a sailor not only had
to do his own job, he had to take the initiative and volunteer for other duties aboard the sub. Some sailors—
the immature ones, Weller believed—thought that was
ass-kissing. But he knew it demonstrated to the petty
officers and chiefs above you that you wanted to be
part of the whole team, that you had the potential to
manage the entire boat.

When Weller became the *Nebraska*'s chief of the
boat two years ago, it had been the biggest management challenge of his life. By a quirk in the rotation
schedules, Weller, Volonino, and the executive officer
then, Lieutenant Commander Duane Ashton, all had
arrived at the *Nebraska* at the same time. That did not
happen often, and for a submarine it could be trau-
matic. Each sub commander runs his boat differently.
And each executive officer and each chief of the boat
has a different operating style as well. If one is replaced, it means there will at least be some continuity
for the crew, because the other two members of the old
management trio will still be aboard. But to have the
sub's top three managers—"the command," as they
were called—replaced all at once means a wholesale

change in the boat's operating style. No crew ever liked that.

Before they walked aboard the *Nebraska*, Weller and Volonino had met, and both decided they wanted to rule more by motivation and consensus than by fear and intimidation. But after they came aboard, they discovered the crew had become far too informal for their tastes. Subs relax military protocol on a patrol, but the *Nebraska* had become too relaxed. Chiefs were calling officers by their first names. Sailors were watching movies during training drills and wearing headphones attached to compact disc players while on the job. One of them had even bopped into Volonino's stateroom with rock music pumping into his ears while Weller was having a meeting with the skipper. Weller about had a heart attack.

The crew felt whipsawed. They thought school was out when the new team arrived. Volonino, Weller, and Ashton were not screamers. But at the same time, these new managers wanted more military bearing aboard the boat. Some sailors liked the fact that Volonino genuinely wanted to include them in the decision-making. Others complained, and still did, that Volonino acted arrogant and aloof and kept his distance from the sailors: Telling them when they could watch movies . . . micromanaging their free time . . . they were being treated like kids, the sailors complained.

Making the new command's fine tuning even more ticklish was the fact that the *Nebraska* was already considered the premier boat on the waterfront. It had won the previous year's "Battle-E," which in peacetime is one of the top honors a sub can receive. The Battle-E award for battle efficiency is given to the best sub in the squadron, based on a complex rating system. Crews competed intensely for it. For this reason the crew wondered why anyone would change the best sub in the squadron.

That Battle-E was ancient history, as far as Volonino was concerned. The *Nebraska* certainly wasn't the best boat *he* had ever seen. It had performed poorly on more recent inspections and the crew was too cocky. The sub had difficult missions ahead of it and Volonino, Ashton, and Weller had to get the men ready for them. They couldn't rest on their laurels.

It took a year of Weller cajoling and haggling with not only the sailors but the chiefs as well, who resisted the new rules. But the headphones came off, some military protocol returned, and movies were banned during drills. Weller could have just issued an edict at the outset, but then the crew would have felt that the command was against them and Weller and Volonino wouldn't get the extra initiative out of the sailors to put in twelve-hour days seven days a week for nearly three months on patrol. With more than a hundred nuclear warheads on board, Weller knew you could have an unhappy crew but not one that was real unhappy.

Of course, he never believed that the unhappiness would reach the point where the men would launch missiles on their own. But the *Nebraska* was a complex machine, and its sailors were technically proficient enough to know the thousands of ways they could seriously damage the ship or put it in mortal danger.

The changing times didn't help. Everyone knew the Cold War was over. Sailors weren't joining the submarine service today to fight the communists. Most had one goal: to get as much education as they could out of the Navy and then get out. Sometimes Weller found it hard to make them understand that the Tridents still had an important strategic mission. He talked constantly about it, not only to the sailors but to their wives, who often questioned why their husbands had to be away for months now that the Russians had given up the fight.

Weller and Volonino had to remind the sailors that the Russians still had nuclear weapons. Rogue nations like North Korea, Iraq, and Iran could rise up with nuclear arsenals. The U.S. nuclear deterrence had to be preserved because it was still a dangerous world out there.

Weller's preaching got mixed reviews. Some sailors thought he was a dinosaur, on the wrong soapbox, fighting an enemy long gone. Particularly around holidays, some would grumble. Was it really necessary for the Navy to send a Trident to sea every two weeks when both the United States and Russia were back home celebrating Christmas? But the lectures had worked with most of the crew. They understood that the United States still faced threats, and they accepted that they still played an important part in its defense. They might not like Weller or the Navy, they might want to get out, but either way they understood why the *Nebraska* was on patrol.

Weller had found over the years that Cold War or no, there was a dynamic to managing sailors confined aboard a sub. Some days you could get them to do anything. Other days they seemed to take three steps backward and nothing could be accomplished. He and Volonino were constantly analyzing crew morale and discussing what was needed to keep the men on track. It took patience. You had to be a psychologist, to know when to push and when to back off. Just as important, he had to have all the chiefs speaking with one voice, behind the command. If the chiefs started commiserating with the sailors, blaming the skipper and the COB, all was lost. The command and the crew would become separated and the sub would do poorly—particularly in the Battle-E competition.

And the Battle-E competition was on Weller's mind a good bit now. The Navy had created various awards that different departments and divisions aboard a Trident competed to win—it helped them keep their fighting edge—but the pinnacle prize for the overall performance of a sub was the Battle-E. Only one boat per squadron received it each year. All kinds of good things came to a crew that won the Battle-E: voyages to the Mediterranean Sea with glamorous port calls to show off the sub, high-profile assignments from the Navy brass, kudos in the military records for the sub's leaders. The trick was that both crews for the Trident had to do well for the sub to win it. William Porter's Gold Crew had to score as high in the competition as Volonino's Blue Crew. It was like two NFL football teams having to have Super Bowl seasons. The crews had to be constantly on edge, the officers and chiefs constantly pushing and training their people.

But like NFL teams, Tridents had good and bad seasons. One year, a sub could be the "hot boat," as the submariners called it. The next year, through no fault of the crew's, it could have bad luck with a major piece of equipment that forced it to keep returning to port for repairs. Like a football team, turnover in key players could hurt. A sub could absorb poor replacements for some of its junior officers or sailors. But if several good department heads transferred out, a new COB or executive officer arrived who didn't work well together, then trouble happened. You couldn't put your finger on it, but typically a boat became bogged down in maintenance problems, jobs took longer to complete, the crew became distracted from its mission. Then the sub began to slip in rankings for the Battle-E. In no time, a hot boat was at the bottom, the grading criteria were so strict.

The Nebraska was now considered the hot boat in the squadron. It had won the Battle-E in 1996 and 1998, and Weller thought it had a good chance of walking away with the prize in 1999. That would be quite a coup. A "three-peat," as he liked to call it, borrowing from basketball slang. But the competition was intense this year. The four other boats in the squadron—the Pennsylvania, Kentucky, Maine, and Louisiana—were gunning for the Nebraska. They thought all that Big Red stuff was too much, and they'd like nothing better than to knock the sub off its pedestal. The crew had had a rough inspection of its engineering department the previous December, which had cost it precious points in the competition, and the Kentucky was an up-and-coming challenger. But Weller still thought the Nebraska was the boat to beat.

And he didn't give a damn if the sub went overboard with the home state pride. The Nebraska was a proud boat and proud of its history. The crew liked being known as the best on the waterfront. They cared about the ship and looked out for one another. There was little of the backstabbing to get ahead that the sailors had found on other subs. Sailors were treated like family. The officers had a cooperative management style, which Volonino set. As a matter of fact, Volonino set the mood for everything on the boat. If he was grumpy under way—and he rarely was—all the officers became gloomy, and so did the chiefs, and it filtered down to the crew.

There was no racial prejudice on board, or at least none that anybody could tell. If there were fissures, they came from the jobs the men performed. The crew was broadly divided into "nukes" and "conners." The nukes were the brainy sailors who stayed mostly in the back tending to the reactor and the rest of the engine room. The conners were the sailors who manned the sub's front end, its cone, where the war-fighting operations took place.

The conners thought the nukes were bratty nerds because they were couped up in the back of the sub most of the time. They even referred direly to "creeping nukism," because most of the officers came from nuclear engineering backgrounds, which supposedly made them unreasonable nitpickers. The nukes thought the conners were dimwits who blindly followed orders. There was a joke that went, "Tell a nuke to do something that's incorrect and he'll question the order. Tell a conner to do it and he'll make the mistake as best he can."

Different departments even got tagged with labels. The sonarmen were considered wise guys, the A-gangers were considered knuckle-draggers, the missile technicians were prima donnas, the navigation technicians were always rubbing shoulders with the captain. None of the stereotypes seemed particularly accurate, and on the *Nebraska*, relations among the cliques was far better than on other boats.

But it was a challenge, nonetheless, keeping them all pointed in the same direction. In doing so, Weller had grown as close to Volonino as he ever would to a senior officer. Stuck underwater for months, the chief of the boat and the skipper had developed a much more intimate working relationship than their counterparts on surface ships. The highest-ranking chief aboard a surface ship is called a command master chief, and he ends up being more the skipper's ceremonial sidekick presiding over functions, whereas on a sub the chief of the boat runs the vessel's day-to-day operations along with the executive officer. Volonino delegated considerable power to all his chiefs—more power than other skippers did, or so the sailors felt—and in terms of responsibilities he considered Weller equal to his executive officer.

It didn't matter how flashy a captain was. Volonino believed that if he had a poor COB, his sub would never be better than average. Weller was the most aggressive and energetic chief he had ever met. He was the first man Volonino talked to as a sounding board before he announced any new policy with the crew, the man who took the crew's temperature for him, the man who could persuade the sailors and chiefs to follow unpopular orders (the crew would gripe that he did this too much and had become the skipper's yes-man), the man Volonino went to first when a department was screwed up.

The two men had been through a lot together running the *Nebraska*. Both had been divorced during the sea tour. Weller had been to Volonino's house many times, where they puffed cigars, sipped beer, and talked for hours about the breakup of their marriages. Weller felt he could walk into Volonino's stateroom at any time to confide or to let out his frustrations. Volonino looked forward to the day when he could be a close personal friend of Weller's after he left the sub. Aboard ship they still had to maintain a superior-subordinate relationship. But Volonino trusted Weller with his life and the lives of the crew.

Weller pulled himself up from his chair to walk back to his office, a tiny cubbyhole next to the crew's study in the rear of the ship. He had more changes in watch schedules to untangle. It was a never-ending chore juggling jobs and sailors aboard the sub. Other chiefs began drifting into the chiefs' quarters to plop on couches and gossip. Weller would find Dillon later. He could appeal to the Supreme Court, but he was still losing the orange hair. Otherwise, Dillon would not have a pleasant voyage, the COB had decided.

6 • Listening

The crew had long settled in for a peaceful night. But four men kept quietly busy in the darkened room no bigger than a narrow walk-in pantry. The only light inside it came from colored dials and the green waterfall glow of the screens on the computers, which constantly beeped and chattered and chirped as if to prove to their human masters that they were hard at work.

The sonar shack was squeezed into a small space just in front of the control center on the port side near the sub's cone. On its front door, which almost always remained shut, a menacing sign warned: STOP: TOP SECRET OPERATIONS IN PROGRESS. Inside the room, Johnson delicately moved the thin joystick with his thumb and forefinger. It was no bigger than a Nintendo joystick, and James Johnson, a nineteen-year-old black man from Cross, South Carolina, about forty miles from Charleston, held it as he would a demitasse cup.

The best music always came late at night, Johnson knew, when the sea life rose to the higher depths where the sub patrolled because the water there had cooled and was richer with oxygen. Ever so slightly, he tilted the joystick to the right. The cursor over the green waterfall on the screen in front of him moved as he did it.

There it was! He stopped the cursor. Underneath it a

series of white dashes fell down the green waterfall. He could barely hear it in his headphones. A high-pitched squeak in the distance.

"Dolphins," Johnson announced. He reached up and turned the volume on the small speaker box so the others could listen. Another sound, this one different, like a little girl crying, could be heard over the speaker.

"Probably a second dolphin," Jason Barrass guessed. Barrass leaned back in his padded metal seat with a bored look on his face. He sat in front of another column of computer screens at the other end of the room. He enjoyed the sea music. "It's probably the only thing I like about the job," he said. "Listening to the biologics."

Sonar picks up a symphony of natural noises in the ocean. Dolphins cry or giggle or screech. Sperm whales sound like workmen hammering nails. The melting and cracking of an iceberg gives off the high-pitched growl a punk rocker makes on his guitar. Shrimp make a snapping sound, like rain falling on a puddle.

Chris Wilhoite, who was in a seat behind Barrass and Johnson, looked up from a three-ringed binder he'd been flipping through and took in the noise. "I was up in the North Atlantic," he began to muse, "when all of a sudden I started to hear this whale screaming. . . . Then it just stopped." He closed the notebook for a moment to remember that sound. "Probably a whaler had just killed it."

All sonar operators become meticulous about detecting the traces of sound coming from their headphones or spotting those white lines etched down the green waterfalls. It takes months and months of listening and watching to make out the difference between

the traces of sound a merchant ship makes and those
that a sub makes. An experienced sonarman can identify more than a hundred man-made or natural sounds
in the ocean. Some of the job is boring, particularly
when the sea is silent. Some of it is exciting, tracking a
mystery noise. Some of it is "PFM," as the *Nebraska*
sonarmen liked to say—pure fucking magic.

Johnson moved the joystick again. The cursor
drifted over another continuous line on the green waterfalls. In the headphones he wore and over the
speaker came a "*clack-clack-clack*" noise. It was a
merchant ship, he could tell immediately.

"Four blades on its screw," Matt Douvres added
matter-of-factly. He was sitting between Barrass and
Johnson. "There seems to be an imperfection in one of
the blades. It's causing a highlight on my screen."
Douvres was older than the others. Born in New York,
he had moved to Los Angeles at seventeen and played
in a rock band when he graduated from high school.
Seven years later, at the age of twenty-five, he joined
the Navy. Since then, he had fallen in love with the
sub's electronics—funny considering that English, not
math, had been his best subject in high school. When
he wasn't on duty in the sonar shack, he tinkered with
the ship's computers as a collateral duty.

From top to bottom, side to side, every inch of space
in the sonar shack was packed with high-tech gear.
Perched on a shelf in the forward part of the room was
a WLR-9 acoustic intercept box that would sound an
alarm if an enemy vessel or torpedo began pinging the
sub with sonar. Lined up along the port bulkhead were
the "stacks," the nickname for the three columns of
BQQ-6 consoles with two computer video displays
stacked one on top of the other in each column.

The consoles displayed and analyzed the noises the sub vacuumed from outside. Johnson sat in front of the broadband stack to the far right. Douvres sat in front of the middle stack, the class stack. Barrass sat to his left in front of the narrowband stack. Then farther left, crammed into the corner where another operator could sit in front of it, was a spectrum analyzer plus the AN/BQR-27, one of the fancy new consoles just installed, whose screen could display sounds farther out.

Wilhoite sat in a seat behind the four operators next to the other new gadget the sonar shack had just acquired, a TAC-4 like the one in the control center whose multicolored screen displayed the course of the sub and other vessels around it. He was the sonar supervisor for this shift. It was considered one of the more important jobs on the sub. Most captains aren't quick to approve a sonar supervisor because it entails such a huge responsibility. Sonar—the word was coined from the phrase "sound navigation and ranging"—looks for noises in the ocean, of which there can be many. The *Nebraska* sailed blind underwater. Once it sank deep, the sea around it was all black. Sonar became both its eyes and ears. Volonino wanted to be able to go to sleep at night assured that his sonar supervisor wouldn't miss a sound that might get them killed.

Wilhoite kept a little green notebook in his breast pocket with a list of reasons why he didn't like the Navy. It was a long list. The sea service had changed too much for his tastes. Too politically correct now. The training seemed more for show than for combat readiness. Too many inspections, too much primping for VIPs the brass was trying to impress who seemed to constantly be coming aboard. The officers could be tight-asses with all their rules about not wearing civil-

ian clothes on patrol or not being too rowdy at parties.
On the fast-attack subs he had sailed, a sonarman
tracked and hunted other ships. On a Trident, he listened for vessels so he could run away from them. The
Nebraska's mission was to stay hidden, not be found
by the enemy so it could fire its ballistic missiles. Not a
very sexy mission for a sonarman.

Wilhoite was married, and his wife, Melissa, was
president of the *Nebraska* wives' club, a tolerable
enough job. Occasionally an officer's wife tried to
wear her husband's rank. But the way the Navy treated
the single guys, particularly the ones who lived in the
barracks on base, was juvenile, he thought, with those
constant, nit-picking inspections. They were petty officers, but the base acted as if they were new recruits,
checking their rooms all the time.

Wilhoite had been on the *Nebraska* for two years
and would be on it for another two. Then he would
probably get out, study for a degree in civil engineering, maybe back in his hometown of Indianapolis.
Melissa was making $45,000 a year as a property man-
ager for an apartment complex. It was time to start
thinking about her career, which got disrupted every
time they had to move to another duty station.

Lieutenant Hardee opened the door to the sonar
shack and popped his head in. He was the officer of the
deck in control now, and he wanted an update on the
contact the sonarmen had been tracking the past half
hour. "Contact" was their slang for any vessel's noise
the shack had detected. In this case, the contact was a
merchant ship puttering along far away from them,
"not doing anything particularly exciting," Wilhoite
told him. Hardee closed the door and walked back to
the control center.

Barrass began writing entries on the touch screen of his computer lap pad, which recorded all the contacts the shack detected and where they were headed. Sonar was all based on the physics of sound, which was all theory, Barrass believed. Maybe that explained why sonarmen were the most argumentative bunch of people he'd ever met. Each of them always thought he had the correct theory.

Sonarmen on the *Nebraska* were young, brash, arrogant, rough around the edges, outspoken, loud, rebellious at times, not afraid to complain if too many extra duties were heaped on them, which happened a lot in port when they were idle and not tracking other vessels. If they had had to identify one common thread among them, they would have said it was that they all had at least one screw loose. They certainly could be obnoxious at times, as they were the first to admit. Wilhoite thought they were the biggest bunch of whiners he'd ever worked with, grown men who'd complain about anything and everything. They hated to lose and loved to brag to other ships that the *Nebraska* had the best sonar shack on the waterfront, which wasn't too far from the truth, considering that they always held their own in competitions, even with fast-attack subs, whose sonarmen were supposed to be more skilled in hunting vessels. In the game of hide-and-seek in the black ocean, the sonarmen put you in the game or take you out of it, they liked to say.

But if they weren't in the Navy, they were sure they would never even associate with one another. The submarine service had forced them to become close friends. There was no alternative. Tough luck if you didn't like another sonarman. You were stuck sitting next to him for months. Holed up in that tiny dark room for six-hour shifts, with the same four guys, could turn them into bored kids packed into the backseat of the family car on a long hot drive with no air-conditioning.

After about a week on patrol, everyone would run out of things to talk about. Then they would tell stories, then they would play the name game, then sports trivia (sonar trivia was outlawed). Then they would start teasing one another, probing to see how far they could go before they got on someone's nerves. You couldn't be thin-skinned in the sonar shack or let weaknesses show. They ate their young. Like family, they could pick on one another mercilessly, but outsiders were not allowed to do it. By the end of the patrol the conversations would degenerate to old whores and the ridiculous.

Using sonar to find the enemy is as much art as science. After the sub's hydrophones detect a sound, the sonarman's job is to pinpoint its exact location so that if it is the enemy making that noise, the sub can hide or, if it has to, fire a torpedo at him. From the raw readings off the screens and through his headphones, a sonarman tries to develop a three-dimensional picture in his mind of where enemy vessels are located.

It is no easy chore. Pinpointing a sound's location, particularly if it comes from miles and miles away, at first involves making assumptions along with a lot of educated guesswork. Imagine being in a large, blacked-out room and hearing a sound. At first you will have only a vague sense of exactly where the noise is coming from. But as you walk around and the sound becomes louder or softer you'll get a better feeling for its location. The same thing happens for a sonar operator. As both his sub and the enemy vessel move around in the ocean, the sound picked up by the sub's hydrophones changes in frequency and intensity. Using sophisticated computers plus a little bit of simple geometry and triangulation, the sonarman continuously refines his estimate of the sound's location until he thinks he can finally pinpoint it.

But like art, location is in the eye of the beholder. Each sonarman in the shack was making his own best guess about the exact spot of a contact. And the guesses never jibed. In fact, Wilhoite would be worried if they did. It was almost impossible to have a perfect solution to an enemy vessel's location. So if all the sonarmen in the shack agreed where a sound came from, particularly if the sound was only recently detected, Wilhoite would be suspicious. His guys probably had all become victims of groupthink and the solution they agreed on was probably wrong.

Finding another sub in the deep black ocean took a lot of sonarmen haggling with one another. They all got a kick out of the movie version of Tom Clancy's novel *Hunt for Red October*, in which a brainy sonarman named Jonesy is up all hours of the day and night single-handedly tracking a Russian sub. In real life, no one is that good. During battle stations when the *Nebraska* played cat and mouse with another vessel, more than a dozen sonarmen would be packed into the shack, all shouting at each other like brokers at the Chicago Mercantile Exchange.

The *Nebraska* had all kinds of ears. Mounted inside its cone was a spherical array, which looked like a giant beach ball with hydrophones around it. The hydrophones vacuumed sounds in front of the sub and, up to a certain angle, on its sides. The spherical array, however, could not hear behind the vessel in what submariners called the "baffle region," because the *Ne-*

braska's own noise masked any other sounds the forward hydrophones might pick up. Therefore, to listen to what was behind it, the *Nebraska* spooled out a cable as long as two thousand feet with a string of small hydrophones at its end. There were also hydrophones arranged along the *Nebraska*'s hull to listen mainly for torpedoes that might home in on the sub.

The noises that all the hydrophones collected were converted into electrical signals and fed to computers that analyzed the sounds and displayed them on the video screens in the sonar shack. Sound was divided into two categories: broadband noise, generally loud or closer to the vessel, and narrowband noise, fainter and farther off.

Think of standing in the middle of a football stadium filled with screaming fans. The noise from the crowd is broadband. Narrowband picks out the cheering of one fan at the fifty-yard line. Broadband is the type of sound given off by a noisy merchant ship that doesn't care who hears it coming. Narrowband is the whisper of another sub trying to be undetectable. A narrowband processor can detect the sound of vessels more than forty miles away.

The two large video displays on the broadband stack that Johnson sat in front of were each divided into smaller screens that showed what looked like hazy green waterfalls tumbling down them. Each of the small screens represented an area of the sea that the hydrophones covered. The white lines that drifted down the waterfalls were visual representations of the sound one of the hydrophones had picked up. The brighter and wider the white line, the louder the noise a hydrophone has detected. Where the line appears on a particular screen and the angle it falls down the waterfall also tells the operator the change in the general direction of the sound—or its "bearing rate," as the submariners call it. By watching how the white line shifted right or left on the screen as both the *Nebraska* and the contact moved, Johnson could begin to narrow down a better fix on the position of the other vessel.

The joystick he pushed moved the electronic cursor across the screens at the top of the waterfall. Wherever the cursor's arrow stopped over the waterfall, Johnson heard the sound the hydrophones were picking up from that direction. The sonar's computers also could electronically track the sound.

With the sounds in the ocean displayed this way on screens, a sonarman relies on his eyes as much as his ears. But the ears are still important. The line on the screen will warn of noise, but trained ears are able to tell, for example, if it is a metallic noise. A sonarman sometimes will close his eyes with the headphones on and appear to be asleep. Actually, he is trying to shut everything else out of his mind to listen for the faintest clue that might identify what is out there.

The video display on the narrowband console, where Barrass sat on the other side of Douvres on the left end of the stacks, was arranged differently. It had narrow, rectangular green screens that displayed specific frequencies of far-off sounds the sonarmen hunted. A white line snaked horizontally across each screen spiking like an electrocardiogram at the points where the hydrophones detected a particular frequency.

Different vessels put out different frequencies of noise from their machinery. Typically, the narrowband operator is the first to pick up a faraway sub, detecting the frequency of sound made by one of its pumps. As a Trident closes the distance with the sub, the broadband operator begins to pick up the vessel and the sonarmen begin to hear noises from other machinery on board, from engine gears that don't mesh perfectly, or from people moving around the boat.

The classification stack that Douvres sat at, between Barrass and Johnson, had video displays with the same green waterfalls that the broadband stack showed. But a "class stack" console does different things with the noise. It electronically demodulates the sound's frequency, much as a car radio does, so it can pluck out tones that identify what kind of ship the sonar has spotted. For example, from the "whuf-whuf-whuf" noise a ship's propeller makes as it turns in the water, the class stack can determine how many propellers a vessel has, how fast they are rotating, and even how many blades each propeller has.

The blade count is important. Merchant ships usually have three or four blades on each propeller. Warships and submarines have five or more. The class stack can also tell if a propeller's revolutions per minute increase or decrease, a tip-off that a ship is speeding up or slowing down.

Douvres accidentally bumped an empty coffee cup on the ledge of his control panel so it tumbled off and shattered on the floor. On the large screen of the spectrum analyzer near the entrance to the shack, a horizontal yellow and red line began floating down its green waterfall. The sub's sensors had picked up noise from the breaking china, and the spectrum analyzer, which was good for deciphering stray noises inside and outside the sub, was now displaying the sound. The sensors could detect the slightest noise the sub might emit into the ocean. Barrass demonstrated by taking a plastic pen and tapping it lightly on a bracket directly mounted to the ship's hull. In the shack, it made a faint tick-ticking sound. Through the headphones connected to the spectrum analyzer, it sounded like a drum tapping, and on the screen a thin yellow and red line drifted down.

There wasn't much else important making noise out there tonight. On most patrols, the biggest threat the sonar shack now listened to ended up being economic. Ninety percent of the ships they now tracked were merchants carrying cars from overseas that were increasing the foreign debt, the operators liked to joke. Rarely did they spot a foreign sub or warship that intelligence reports hadn't already warned them was sailing in the area. In fact, far out in the lonely ocean they could go for weeks without detecting another vessel.

The shack had been a far more lively place to work during the Cold War, when the *Nebraska* could almost always expect two or three nearly close encounters with a Soviet attack sub trying to locate it during a patrol in the North Sea. Nearly close, because the Soviet hunters never found an American Trident.

But those days were long gone. Practically the only times the sonarmen played hide-and-seek games now were during training exercises when other American subs tried to find the Tridents. The American subs used Russian tactics when they hunted to keep the Tridents trained for the old threat. Hell, we might as well phone the Russians and arrange to play war with them so both sides could stay in shape, Wilhoite would say only half jokingly.

No one would ever accuse him of being a fan of the Navy, but Barrass still thought the Tridents should remain on patrol. He believed that deterrence was still needed. The Russians still had nuclear-tipped missiles they could fire at the United States. There were other enemies out there like Iran. Who knew, one day one of the rogue states might dispatch nuclear terrorists to strike.

Barrass had a buzz haircut that made his ears look even bigger than they were. His front teeth lapped slightly over his bottom lip. It made him look like a skinny towheaded teenager. But Barrass was twenty-two and already he had been through a lot in life. It seemed to him that he'd spent most of his growing-up years in one seedy neighborhood after another in south Florida. His parents had been divorced when he was three, and his mother, who cared for him, drifted from job to job every couple of years. He spent his last two years at a high school in Kissimmee, Florida, near Disney World, where his mother was working as a hostess in a sports bar. Barrass had decent enough grades to win a four year scholarship from the Air Force, which was the only way he would ever attend college, because his mother never made enough money to pay for it. By the time he graduated from high school she had moved up north to another job.

Barrass enrolled in the University of Central Florida, where the Air Force wanted him to study nursing. One trip to the hospital convinced him that wasn't the career he wanted, so he dropped the Air Force scholarship and tried to stay at Central Florida with the little financial aid he could scrape together on his own and money from a part-time job.

He lasted only two years. By then he was living with Amanda, who would later become his wife. They had met in high school and together had gone to Central Florida, neither receiving any help from parents. They were on their own at the age of nineteen. Barrass's

mother by then was living in Chicago with her third
husband. Jason and Amanda both worked forty hours a
week at Disney World, an hour's commute each way
from their apartment off campus. They could pay for
their tuition, but there was never enough money left
over for food and the rent. It had been an awfully
young age to have so much responsibility heaped on
you. They both had grown up quickly.

. Barrass decided to join the Navy in 1995 to earn
money for college. He found the military regime bearable as long as he kept his mouth shut and didn't volunteer. The schoolwork was easy—he had always been
good at quickly comprehending subjects—and he finished first in his class at boot camp and submarine
school, then second in sonar school. Amanda and he
arrived at Kings Bay in December 1996, married by
then, with Amanda pregnant and due in a month.

That was when Barrass got his first dose of reality
with submarine life. The *Nebraska* was about to set
sail, and Barrass's pleas to remain behind until his wife
delivered fell on deaf ears. No matter that Amanda's
pregnancy had been difficult. You go to sea, he was
firmly told.

The patrol had been unbearable for him. He studied
his qualifications for his dolphins like a madman to try
to keep his mind off Amanda, who he knew would deliver their first child alone and miserable in Kings Bay.
For almost a month, he heard not a word from shore.
Amanda, it turned out, had had a grueling forty-seven
hours in labor. Early on the morning of January 5, the
commander who skippered the *Nebraska* before
Volonino had the sub rise to periscope depth and ordered Barrass to report to the conn. He swiveled the
periscope around so its lens pointed to the southeast-

ern coast of the United States. The commander had just received a radio message telling him that Amanda had delivered a nine-pound, ten-ounce boy the night before.

"Here, take a look at the sunrise on your son's first day," the commander said gruffly, pointing to the scope. It was the nicest thing anybody had done for him up to that point in the Navy.

Even so, Barrass had a low opinion of the officers and chiefs, who he thought lorded over the sub like plantation foremen. As far as he was concerned, they cluttered the workday with too many drills and micromanaged his free time with too many nit-picking rules about when he could watch a movie or listen to CDs. All these awards the sub received, like the Battle-E, were so the officers could polish their records. He didn't give a damn about them.

He'd go nuts if he wasn't in the sonar division. The sonar operators and the radiomen were the only ones on board who had any contact with the outside world. Those poor souls back in the engineering division, locked up in the engine room for months, which always smelled to him when he went back there like a heavily chlorinated pool, had no idea where they were on earth.

As it was, Barrass had done well aboard the *Nebraska*. Volonino had promoted him quickly to petty officer second class, because he had worked hard as the ship's oceanographer. Barrass had found that collateral duty more rewarding than anything else he had done on the boat. After he took the position, the *Nebraska*, which had been scoring below average on oceanography during inspections, started making above-average grades.

Oceanography had become an important discipline for American subs. Until the early 1980s, U.S. sonarmen had comparatively little trouble finding and tracking Soviet subs because they sailed so noisily. But after Navy turncoat John Walker sold precious secrets to the KGB, Russian subs learned to hide better from their American hunters. The U.S. Navy had to look for other advantages, like the ocean's environment, which the Russian subs also began using more to hide. Barrass now kept up a computer program packed with data on the contours of the ocean's bottom, the composition of the water, its fronts and eddies and gulf streams. Sound travels differently through different contours of the ocean bottom, which affects how close a Trident can be to an enemy sub before being detected. He developed an oceanographic picture and overlaid it on the navigation charts to show how the *Nebraska* and enemy subs could hide in the underwater environment.

This was Barrass's fifth patrol on the *Nebraska* and his last. His enlistment was up in October, and he had enough money saved to enroll in the University of South Florida at Tampa. He would again join the Air Force ROTC program and this time see if he could be a pilot or intelligence officer. The Navy had dangled all kinds of cash in front of him to reenlist, but he was ready for this part of his life to end. When he became an officer, Barrass promised himself, he wouldn't turn his nose up at the enlisted ranks. He'd be a better leader because he wouldn't forget his roots. He'd treat his men with respect.

The sound-powered phone in the shack rang. Hardee was on the line. It had been an hour since he dropped by sonar and he wanted another update on the contacts.

There were two out there, Wilhoite told him. One was

a merchant about twelve miles away. The identity of the other was still a mystery, and sonar didn't yet have an accurate fix on where the strange sound originated.

Wilhoite hung up the phone and peeked over Barrass's shoulder at the narrowband screens. "It looks like a three-blader," he said, eyeing the peaks in the horizontal lines. "Probably a merchant."

"I don't know," Barrass said. "It sounds like environment . . . unless it's a really quiet contact."

Wilhoite grabbed a headphone off the panel. "Plug me in," he said to Johnson, who was sitting at the class stack.

Wilhoite listened for a moment. "Go through your filters," he told Johnson, who began pushing buttons so the noise they were interested in could be heard better.

"It sounds like loud environment or a quiet contact," Barrass still insisted.

"I'd track it," Wilhoite decided. The shack's computers would automatically follow the sound. "We're going to call it a merchant. Probably a trawler with biologics around it." Sometimes the natural racket could be a problem, particularly when sea life swarmed around an enemy warship or sub dumping trash. The chatter from the fish and shrimp drowned out noise sonar would normally hear from the vessel's machines. Every afternoon at the four-o'clock briefing for Volonino, a sonar supervisor would describe the background noise he expected in the waters in which the sub sailed. The sub also routinely shot out a submersible bathometer in a small canister, which sank to the bottom of the ocean and gave what amounted to an underwater weather report.

Wilhoite walked out of the shack to the control center to report to Hardee that sonar for the moment thought the second contact was a merchant.

"Rotate," Douvres announced. The three sonar operators stood up and switched seats. They played musical chairs every hour so they wouldn't develop tunnel vision from watching just one console. It could be mesmerizing staring at the waterfalls. Sometimes Wilhoite, because he was scanning all the screens, could pick up changes on the screens faster than the sonarmen sitting in front of them.

Barrass moved to the broadband stack, Johnson to the class stack. Douvres plopped in front of the narrowband stack.

Wilhoite returned to the shack. "Keep analyzing it," he told the three operators. The mystery contact probably wasn't a sub. But it might not be a merchant trawler. "Intel reports that a U.S. warship may be working in our area. That may be what we've got."

Wilhoite headed out the door again for the navigation department, which kept records of Navy ships on patrol. "I'll be back," he said over his shoulder. "I'm going to find out what kind of warship it is."

Douvres stood up to make a run to the galley two floors below to fetch drinks for the others.

Wilhoite returned a minute later. "It's a fast frigate," he reported, the smallest warship in the Navy's fleet.

"Yeah, it's definitely a surface contact," Barrass said, squinting at his screens. He'd given up on the environment theory. The environment he'd thought he'd seen and heard was probably heavy wind or rainfall over the sea, which always caused a racket.

Hardee phoned again. "It might be a warship," Wilhoite told him, but he wasn't ready yet to call it. A trawler could also be making that noise. They'd need to analyze it more.

In the control room, Hardee buzzed the skipper. Volonino had ordered that he always be informed when the sub detected another warship, even if it was a U.S. warship.

The three sonarmen stared intently at the screens, with headphones covering their ears. This was the fun part, zeroing in on a mystery noise. Sounds became their passion. Sitting for hours listening to the sea gave them all an acute appreciation for sounds when they returned home. The chirping of birds, the honking of car horns—sonarmen became far more sensitive to them than others back on land.

Douvres returned with a tray of sodas. Barrass now began to suspect as well that the mystery contact was a warship. He could see faint signs on the class stack screens Johnson was watching that the contact had two screws rotating two propellers. Few merchants had two screws. Warships usually did.

It looked like the warship the intelligence reports said was in the region, Barrass and Wilhoite decided. Another giveaway: the contact seemed to be moving erratically. A merchant usually chugged along more slowly, keeping to a predictable course and speed. Warships often zigged and zagged so enemy snoopers couldn't detect a pattern to their sailing.

A messenger walked into the shack with a copy of tomorrow's plan of the day plus the menu for tomorrow's meals. Seafood Newburg for lunch. They all groaned. "Nothing they make on this boat is seafood," Barrass grumbled. "Seafood Newburg sounds good, but don't believe it."

Barrass kept shifting the cursor on his broadband screens to the left. He still wasn't completely sure the second contact was the frigate. But the white line on the screens kept on bending, indicating the contact still was changing speeds.

If it was a U.S. warship, "he doesn't have a clue that we're out here," said Wilhoite. "They're probably playing cards on the ship," oblivious that a Trident was watching them. He was sure there was no chance the frigate would detect the sub.

"He's about to break," Barrass suddenly said. It looked as if the contact was shifting course again and heading toward the *Nebraska*.

For another minute, Barrass watched the screen intently. "He's breaking," Barrass finally said. "The warship is close."

Wilhoite dashed into the control center to alert Hardee.

Barrass was somewhat perplexed. "He's going to fly right by us and yet still no screw blade," he said. By now the sonarmen were almost certain it was a frigate. But they should have also had clearer indications on their screens that it had two screws.

Barrass's best guess was that the warship was eight thousand yards away. That was more than four miles. But the *Nebraska* crew considered 8,000 yards to be close, and Volonino had ordered that the sub stay much farther away from other ships when it was submerged.

"He's still not giving us a screw blade," Barrass said quizzically. He leaned forward in his seat and strained to listen to the scratchy noises coming through his headphones.

"He's going to light off," Barrass finally said warily. Warships and fast-attack submarines can ping the ocean with an active sonar wave to see what the sound bounces off of in order to gain a fix on a contact. At this point, if the frigate went active with its sonar, it would be shooting in the dark, because it had no idea that someone was watching.

But if it did it at this range "that would be very inopportune for us," Wilhoite said dryly. "It would light us up like a Christmas tree."

Barrass didn't think the frigate would start pinging out of the blue. Why should it? The frigate's crew thought it was alone in the ocean.

"And we still haven't gotten the screw blades?" Wilhoite asked again.

"I don't know," Barrass said, reconsidering. "It looks like a doublet to me."

The contact shifted course again. The warship was now ten thousand yards away. "He opened up the distance quickly there," Wilhoite said, intrigued. Maybe their first guess that the ship was 8,000 yards away had been wrong.

"He's starting to sound weird," Barrass said and reached up to turn on the shack's speaker so Wilhoite could listen to the static without the headphones on.

Nothing distinctive about that, Barrass decided, and turned down the volume.

Wilhoite began tuning the spectrum analyzer to see if it offered any clues. Then he grabbed a six-inch-thick manual with SECRET stamped on it that identified ships by the sounds they made. Every vessel in the ocean had its own acoustic signature. The sonar shack could fill a refrigerator with the classified manuals describing the sound frequencies emitted by thousands of U.S. and foreign vessels. The manuals had secret reports on the machinery noise that could be expected from each ship or sub, its speed and turning capabilities in the ocean, the areas of the ocean each sailed, the operating procedures of its crew, sometimes even the

work habits of its captain. The shack also had a classified sonar search plan for the parts of the ocean the *Nebraska* would sail, which detailed the types of ships,
friendly or unfriendly, that the sub could expect to encounter so the sonarmen could search for the frequencies they emitted and identify them.

All the detective work made them thirsty again, so
Johnson was dispatched for more drinks.

Wilhoite was now sure it was a U.S. frigate. But
none of the readings they were seeing on the screens
matched the frequencies listed in the manual.

Whatever. In fifteen minutes it didn't matter. The
frigate was sailing even farther away from the sub now
and was no longer of interest to the shack. Johnson
phoned from the galley. He'd forgotten the drink orders. They gave them to him again and made a note to
harass him about it later.

Todd Snyder dropped by to check on contacts. Snyder was the chief in charge of the sonar division. He
had been in the Navy fourteen years and had only recently been promoted to chief. The sailors in the shack
were still adjusting to Snyder's wearing khakis when
just months before he had been a petty officer like
them. But Snyder was already making his reputation
on the boat as one of its most high-energy chiefs.

"No contacts now," Wilhoite told him. The merchant
ship had long gone away and the frigate was no longer
a concern.

Johnson, who had returned with the sodas, nevertheless logged the information on the two contacts into
the computer lap pad. The sonarmen began talking
about politics. It was still early in the patrol.

7 • Casualties

David Smith spun the green wheel to shut the valve that controlled the seawater rushing through the pipe between CG-2 and CG-4. He then rushed to the phone for the sub's speaker system.

"Flooding in the torpedo room!" he shouted into the handheld mike. For submariners throughout the ages, flooding anywhere has been the most terrifying warning that could be sounded. There is no greater horror the men of a Trident could imagine—save perhaps for launching the missiles—than being trapped under the ocean with water rushing in like earth covering their grave.

Alarms began blaring throughout the *Nebraska*. Crewmen scrambled to their stations. The control center made preparations to shoot the sub to the surface if necessary, while officers and sailors slid quickly down ladders to the *Nebraska*'s bottom level near the bow where the torpedoes were stored and fired. In addition to the missiles, the Trident kept Mark-48 torpedoes aboard—each with three times the explosive power of the 1995 blast that leveled the Murrah Federal Building in Oklahoma City—just in case it had to fight other ships and subs. A thirty-year-old petty officer second class, Smith was one of the sub's most experienced tor-

pedomen. He had been trained to react quickly to any
danger of flooding. What a sailor did in those first few
seconds of an accident usually determined whether the
sub would recover from it.

In this case, Smith happened to be standing watch in
the torpedo room when Shawn Brown started waving a
green rag. This was not a real emergency. It was Friday
morning, May 7, the beginning of another day of simulated disasters. The *Nebraska* crew held endless drills
to test response to practically every part of the sub
leaking. Back at Kings Bay, the Navy had a two-story-
high training room replicating a part of a Trident sub,
where large pipes would suddenly burst and water
would rush in as if from a fire hydrant. Trainees had to
shut valves and patch pipes to keep the water from filling up the room. It was too dangerous for a submerged
sub to have real water pouring in for a training drill, so
to simulate such an emergency Brown waved a green
rag at a pipe, which the crew knew was the signal for
them to respond. A green rag meant that seawater was
leaking into the sub from the outside. If Brown had
been waving a blue rag, it meant fresh water was leaking from inside the sub. A white rag represented a
steam leak.

Practically every submarine movie that has been
made has a scene in which hapless sailors are trapped
in a flooded compartment with water rising to their
necks, then over their heads. The films always unnerve
real-life submariners. The scene, however, is unlikely
to happen in modern Tridents. The sailors probably
would not die from drowning. Unlike its ancestors, a
modern Trident does not have many small compartments that could fill up to the ceiling. Rather, it is divided into a few large open spaces. If those spaces

flooded, long before the sailors could drown, the water's extra weight would sink the sub to the ocean bottom, where, in the deep parts of the ocean, the outside pressure would collapse the hull. The crew, perhaps mercifully, would perish instantly from the implosion's shock rather than suffer a slow death from drowning.

But unless a sub collides violently with another vessel or is severely damaged in wartime by a torpedo or depth charge, major flooding casualties are extremely rare. But they can still happen. On August 12, 2000, the Russian nuclear-powered submarine *Kursk* sank to the bottom of the Barents Sea after an accidental explosion in its forward torpedo section blew a gaping hole in its hull. Water flooded in and killed all 118 crewmen aboard.

Modern subs like the Trident are designed with elaborate flood control and hydraulic systems. A labyrinth of pumps and valves for the sub's internal pipes can be opened or closed to stop internal water flows. Hull valves can also be shut to prevent seawater from coming in. Nowadays, the worst flooding that occurs on a U.S. sub is more a controlled leak, with water spraying out from a corroded pipe or a broken valve—faster than, say, from a burst pipe in a home, but not at a fast enough rate to sink the ship. The *Nebraska* was only six years old, so its piping system was still in good condition; the only leaks the sub had suffered so far were small ones at spots where corrosion had formed.

The first-response team hopped down the ladder into the torpedo room. The team included Shafer, who was carrying the team's large canvas bag crammed with wrenches, pliers, hacksaws, patches, a band-it kit, and assorted wooden plugs and wedges. Bob Lewis,

who had just come aboard the *Nebraska* as its new
chief of the A-gangers, led the team. The simulated
leak was a small one that had sprayed a thin stream of
water from a crack in a pipe under the room's left rack
of torpedoes.

Lewis knew exactly what to do. Before arriving at
the *Nebraska*, he had been an instructor at the Kings
Bay base's flooding and fire trainers. "Let's put a soft
patch on it," he said, turning to Don Lee, an A-ganger
he supervised. If it had been a bigger leak with water
pouring out under more pressure, Lewis would have
used the band-it device to clamp a metal patch to the
hole with steel bands. Some leaks, such as a rupture in
one of the pipes carrying steam in the reactor section,
could be dangerous to fix. The hot steam billowing out
could cook a man in minutes. Sailors therefore donned
insulated silver suits with hoods and cool air piped into
them, which could withstand five hundred degrees of
heat for about fifteen minutes—enough time, with
luck, to identify the leak and shut the valve to cut off
the steam flow. For a minor leak like this one, Lee
crawled under the torpedo rack and slapped a rubber-
like plastic patch on the hole. Lewis crawled behind
him with a marline, a jute cord impregnated with tar,
which he carefully began wrapping around the pipe to
seal the patch over the hole.

Alan Boyd climbed into the torpedo room just seconds after Lewis's team. Hardee followed close behind
with a sound-powered telephone slung around his neck
and a cord that could be plugged in nearby so he could
relay messages to an emergency command center set
up in the officers' study to monitor the casualty. The
crew called this "damage control central." The procedure was always the same for these emergencies. The

executive officer rushed to the trouble spot to take personal charge of the crew's response, while Volonino remained at the conn to drive the sub.

Boyd took a quick look at Lee and Lewis lying under the torpedo rack, then turned to Hardee. "No injured personnel," he said crisply. "Damage limited to wetted equipment."

Hardee quickly passed the report along to damage control central. One important aspect of fighting any casualty on the sub was relaying information on the problem quickly so everyone stayed on the same sheet of music in fighting it. In this case, DC central needed to know as soon as possible that the flooding was under control so the sub would not have to broach the surface and expose itself to the rest of the world.

Boyd took another look under the torpedo rack. "The crack is between CG-two and -four," he said, turning around again to Hardee. "Have the wipe-down team come to the torpedo room." This would be another group of sailors with rubber gloves, rags, and towels to begin mopping up the water that had soaked the equipment. Hardee relayed the report.

Boyd issued his orders matter-of-factly. This was a routine drill. The crew could probably shave more seconds off its response time, but the teams had arrived quickly enough so that if it had been a real leak there would have been no serious threat to the sub.

Boyd had no complaints. He wouldn't have been shy about voicing them if he had. One of Boyd's biggest jobs was making the trains run on time in this operation. He oversaw the ship's schedule—called the "plan of the day." Without a plan of the day, the sub and crew were lost. Practically every hour of every day was filled with training or jobs that had to be performed.

The crew was still getting used to Boyd's way of enforcing the plan of the day. He had come to the *Nebraska* two months ago with a long list of ideas for how a sub should be run, and now that he was an executive officer he was in a position to try them out. But not all at once. The crew could take only so much change so fast. And he was still scoping out Volonino.

Boyd had been on a career track reserved for the submarine service's best and brightest. He had excelled in math and science at a private school in Chicago and had picked the Naval Academy over Princeton University because he thought it offered a stronger undergraduate science program. At Annapolis he had been one of only a dozen in his class of about a thousand to major in chemistry and had graduated as a Trident Scholar (his honors project for that award had been to develop a computer simulation for conventional explosives detonations).

Boyd had landed plum jobs in the submarine service—a tour with the prestigious Naval Propulsion Program, then by age twenty-eight the engineer aboard a fast-attack sub—and had earned a master's degree in engineering administration along the way. Before arriving at the *Nebraska* he had been an inspector for the Atlantic Fleet's Nuclear Propulsion Examining Board, the "dark side of the force," as its officers are called, because they are the ones who rule whether the crews of nuclear-powered aircraft carriers, surface ships, and subs pass or fail inspections on the safe handling of their reactors. He knew the back end of a Trident cold.

It is almost more difficult to be selected as a sub's executive officer than as its skipper, the competition is so intense. "Now you're on the varsity team," his personnel officer had told him when he phoned him with the news that he'd been picked. "You've actually made it." Succeed in this tour and a sub command would likely follow.

Boyd knew the *Nebraska* was a hot boat before he arrived. He had inspected it once when he was on the dark side and had been impressed. Still, there was room for improvement. There were procedures he wanted to change, new ways of operating the reactor more efficiently, for example, that he had learned during his inspection tours.

But it was Volonino's boat. Navy regulations give few specifics about an executive officer's duties on a sub. The captain spells them out and decides how much leeway his second-in-command has. Volonino was the king of this boat. Boyd knew that.

But Boyd was the crown prince. He knew that, also. And he had no intention of being timid. He was the first to admit that he had a brusque personality, and he didn't apologize for it. Sure, he would have to adjust to the way the *Nebraska* did business, but they would also have to adjust to him.

Petty officers had quickly discovered that the worst answer they could give him when he questioned a procedure was "That's the way we've always done it." Boyd had begun tinkering not only with the reactor operation, but with how things were run in the front of the boat, in the navigation and weapons departments. From day one, he kicked back paperwork with grammatical errors and began revamping how sailors were evaluated for promotion. Many of the chiefs found him abrasive. And even before the *Nebraska* set sail, he had begun snapping at officers who arrived at meetings late, one of his pet peeves. An officer tardy just five minutes meant that the ten other people in the room

lost five minutes each twiddling their thumbs waiting
for him. That was fifty minutes combined of wasted
time, by his calculation. Condone tardiness and days
could be wasted on a patrol. One of the reasons Boyd
had snubbed Princeton was that its recruiter had been
late for their appointment—twenty-four and a half
hours late, to be exact.

Volonino had no problem with Boyd's being the
heavy. In fact, he wanted it. Before Boyd arrived at the
Nebraska, Volonino had sent him a long letter detailing
what he expected of his executive officer. Volonino
liked to put things on paper. In fact, before he took
over the *Nebraska,* he had spent months drafting a fif-
teen-page, single-spaced statement of his "command
philosophy" and what he expected from the crew,
copies of which he distributed to every sailor, chief,
and officer on board. All skippers wrote command philosophy statements when they arrived, but most tried to
keep them from being too long or detailed so sea
lawyers on board wouldn't later hold them to every
clause and comma.

In his letter to Boyd, Volonino saw himself as the
coach of the sub, the fatherly type, the one the crew
looked to for support and inspiration. Boyd would be
the "enforcer," the stickler on standards, the man the
enlisted men and other officers found impossible to
satisfy, with the little body and big head and bared
teeth like Rickover, demanding perfection. If liberty
had to be canceled, the executive officer announced it
over the loudspeaker. If the sub planned to dock at port
early, the captain announced it.

There is a mystique that surrounds a ship's captain,
and Volonino intended to preserve it. The crew wants
to see its skipper out and about, praising them and pre-

senting them with awards. It is not healthy for the captain always to be the disciplinarian or critical of the crew. Better for the executive officer to have that role. Sailors recover from a harsh word from the XO better than they do from one from the CO. Boyd's predecessor, Duane Ashton, had been uncomfortable being so demanding, and Volonino had had to prod him to sharpen his edges. Boyd had no problems being the bad cop.

Volonino also didn't think Boyd would need any prodding. He knew that Boyd was still in his honeymoon phase, not only with his new skipper but also with the rest of the crew. The crewmen would cut the new XO a lot of slack the first month on patrol. Then they would be more demanding of him. They would test him on how much of an asshole he was willing to be, and Volonino would know then better if he and his number two made a good fit.

The dynamic between the captain of a Trident sub and his executive officer is important. Volonino had served on subs where the CO and XO clashed, where there were petty power struggles. These boats always performed poorly.

The friction never would reach the point of mutiny, Volonino felt sure. Practically every Trident submariner had seen *Crimson Tide* and been jarred by it. In the 1995 movie, Gene Hackman plays the autocratic captain of a Trident and Denzel Washington plays his rebellious executive officer. The two men fight over whether they have been ordered by their high command to launch nuclear weapons against Russia during a crisis. They have gotten a message to launch, but then a radio that is receiving a follow-up message is knocked out of action. The garbled transmission that comes through hints that the first order has been rescinded. Hackman wants to fire off the missiles based on the first order, but Washington has qualms and leads a mutiny to block the firing, saving the world from accidental nuclear war. The second order has in fact rescinded the first.

Trident submarine officers still discuss *Crimson Tide* during private seminars on commanding a ship. But neither Volonino nor Boyd believed that kind of plot could ever unfold in real life. For one thing, the *Nebraska* had more than two dozen radio receivers to retrieve messages from land—there was backup even if the entire radio shack burned down. Plus there was a stack of top-secret manuals on board with contingency plans Volonino and Boyd had agreed on ahead of time for dealing with garbled or incomplete launch orders. For another, it would be hard to find an executive officer in the U.S. Navy who has any qualms about launching nuclear weapons if ordered. Trident submarine officers are screened formally and informally throughout their careers on the question of whether they would carry out a strike order from the President. The hesitant ones are weeded out quickly. The ones left are all like-minded by the time they reach command ranks.

Besides, Volonino would never let relations with his executive officer be poisoned to the point that the two might fight over the use of nuclear weapons. He'd fire him long before that. He would march into the squadron commander's office and demand that he have the XO yanked off the ship. Trident captains are powerful men in the Navy. Boyd, like every other executive officer, knew that he wouldn't ascend to the rank of commanding officer through mutiny. The only way

Boyd could move up was if the duo now succeeded. There was no organizational incentive for an XO to fight his CO.

Boyd was now preoccupied with cleaning up what was left of the flooding emergency. The wipe-down team stayed busy simulating the mopping of the deck. But Brown announced that the leak had also flooded the bilge underneath.

"Recommend pumping water out of the bilge," Boyd said to Hardee, who passed it on to damage control central. DC central gave permission for the pumping.

Boyd stooped down for another look under the torpedo rack. Lewis had finished wrapping the cord around the plastic patch.

With the skipper's approval phoned down from the control center, Smith then spun the green wheel to open the valve that would let water flow through the pipe once more. Brown ruled that the soft patch fit snugly enough that no water would drip out. If it had been a real leak, they would have left the soft patch on and kept the water running.

"The controlled leak has stopped with the installation of a soft patch," Boyd announced, and Hardee again passed the information on his mike.

Brown had seen enough. "Secure from the drill," he ordered.

Brent Kinman scurried about in the lower-level missile compartment like a Hollywood director about to shoot a big scene. Brown and Steve Dille, a senior chief who supervised the engineering department, plunked down boxes full of gray blankets and green rags, plus one beacon light, and began arranging the blankets at the space between missile tubes nine and ten. The men spoke in low voices, trying as best they could not to arouse too much attention from the rest of the crew.

They crouched down in the narrow center passageway. The wide missile tubes were painted burnt orange; piping that snaked around them was painted the same color, and gray boxes were attached to each. Beside every tube stood a barrel-sized gray canister used for creating the high pressure needed to propel the rocket out of the sub.

Kinman and Brown were about to start a fire. In this case, it was a fake one. Every month, the drill coordinators had to stage a major fire among the missiles to test how well the crew could combat it. While flooding casualties are rare among the Navy's submarines, fires are not. Every sub, including the *Nebraska*, experiences them at one time or another. Most are minor, caused by an electrical short or a smoldering cigarette tossed in a trash can in the smoking section. They are put out within minutes of being started. Tridents, for example, have sophisticated sensors in the engine room that can warn when electricity is arcing over a broken circuit and threatening to ignite material nearby. Usually the easiest way to halt an electrical fire is to flip off the circuit breaker to stop the electricity from flowing to the equipment that is burning.

But there is always the danger of other materials aboard the sub igniting that could be far more difficult to battle. The galley's deep-fat fryer could light up, producing thick black smoke. Oil fires from leaking pipes can be ugly. Metals that catch fire, such as magnesium, take forever to extinguish. The hull's insulation is fire-retardant but not fireproof, and flames there could soon encase the sub in cooking heat and hydrogen cyanide gases from the smoke.

Out-of-control fires can become horrifying for the crew. A blaze in a confined space builds up air pressure to unbearable levels. Dense smoke can fill the room, blinding men, then quickly asphyxiating them. Within a minute, temperatures can rise to a thousand degrees (skin melts at 260 degrees).

Because of the small number of crewmen aboard, everyone on the *Nebraska* had to know how to fight fires. Volonino wanted to send every sailor, chief, and officer through the advanced fire-fighting course back at Kings Bay. In a fire trainer there, propane tanks created flames that engulfed a room and students clad in heavy protective gear charged in with extinguishers and hoses to practice battling the blaze. It was like stepping into hell. If nothing else, the instructors hoped the trainer would help the men overcome the natural fear humans have of fire so their first instinct would not be to run away, but rather to attack the blaze to save the ship.

Kinman stooped down for a last look at where Dille had placed the beacon light between tubes nine and ten, then stood up and straightened his back.

"All set," he said to himself, bouncing from one foot to the other and rubbing his hands as he always did. Of course, it would be too dangerous to stage a real fire in the sub, so the drill coordinators again used props to simulate it. The flashing red light from the beacon would simulate the flames. Dille and another petty officer would wave the gray wool blankets to simulate the heavy smoke being produced. The green rags would be spread out to simulate water collecting on the decks and in the bilges as the hose men battled the blaze.

No chief or officer on the sub had ever seen a fire

like the one they were about to stage. A blaze in the missile compartment could be extremely dangerous. Hot, toxic gases would billow out. High explosives might ignite. The missile might leak deadly tritium. But this was the worst case. Never in the history of the U.S. Navy had a Trident crew ever come close to experiencing a fire casualty this severe. The missiles were heavily protected in their tubes and the solid fuel in the rockets was extremely stable and difficult to ignite accidentally. The *Nebraska* crew nevertheless trained often to combat it.

Everyone was in place. "Go," Brown whispered into his pen mike, and turned to Dille, who was crouching between the missile tubes. "Start it."

Dille turned on the beacon light, stood up, then began waving a gray blanket over it. A Hoosier, Dille was a workaholic, meticulous to a fault about every job he was assigned aboard the sub. The officers were sure he would be a master chief one day. Thin, spectacled, and determined-looking, he had become a leader among the chiefs, though he was only thirty-six years old. He was one of the men Volonino depended on to keep the entire sub in sync.

Greg Murphy had the roving patrol for this guard watch. Looking after "the house"—the nickname the sailors gave the missile compartment—was a breeze. Rarely did anything go wrong with these rockets. But as he turned the corner to the lower-level missile passageway, swinging his billy club, Murphy saw the flashing beacon. He groaned to himself—another drill—but only for an instant. In the next second, he reached for the phone that hung in a cradle attached to tube nine in order to alert the control center, then he raced to the forward end of the passageway to grab a fire extinguisher clipped to a bulkhead. He came back to the beacon with the extinguisher's hose pointed at the deck where it flashed.

Brown told him a fire extinguisher, for this drill, wouldn't be enough to put out the blaze. Dille raised the gray blanket and continued waving it over the flashing light to simulate the blaze spreading with more smoke pouring out.

Murphy ran back to the forward section to let control know that the extinguisher hadn't put out the flames.

"Missile emergency; fire in the missile compartment!" the sub's speaker system announced loudly. The missile emergency and fire alarms sounded— "*bong, bong, bong, bong, bong*." Throughout the sub, watertight doors and hatches slammed shut.

In the control center on the upper level, the officers and enlisted men rolled down their sleeves, buttoned up the collars of their poopie suits, and grabbed breathing masks from overhead pouches. Al Brady, who was now the officer of the deck, had immediately ordered the diving officer to bring the sub to a shallower depth after Murphy's first call, so it would be ready to go up even farther, to periscope depth, if the boat had to be ventilated. Brady could always sense when the crew was about to be hit with a simulated emergency. People wearing red drill coordinator caps always started buzzing around control just before all hell broke loose.

Sean McCue, a chief in the missile division who was the chief of the watch at this point and was sitting before the ship's control station, had already flipped two switches on the ballast control panel to pressurize fire hoses with seawater. The *Nebraska* would never run out of water to battle fires, because it came from the ocean. A trim tank could be flooded and water pumped from it to the hoses. The only danger was filling up the inside of the boat. Using a printed matrix with the location of the nearest hoses, McCue began calmly directing fire-fighting teams over the loudspeaker system to head to the missile compartment. He knew it was important not to show any panic over the speaker system. The fire would create enough chaos by itself.

The officers manning the control center wanted to take a look around before they poked a periscope out of the water. They would have more to worry about than the fire if the ascending sub banged into the bottom of a ship's hull, or, in wartime, rose into a nest of enemy vessels on the surface. Sonar searched for ships in the area. Because the towed sonar array wasn't spooled out, Brady conducted a "baffle clear." The *Nebraska* spun around to the right so the sonar sphere in the bow could listen for ships or subs that might be behind the vessel. There were none this time.

The three-man hose teams coming down each end of the narrow passageway looked like space monks doing a duck walk. The seamen wore black oxygen masks with wide visors plastered to their faces, along with thick leather gloves for their hands. Fire-resistant brown cloth hoods covered heads, necks, and the top part of their shoulders. The breathing masks they wore, called EABs (for emergency air breathing), had long, thin rubber hoses they could plug into outlets all over the sub to inhale air. The outlets were painted bright orange, and on the deck underneath each of them were raised, textured triangles so that the men could feel their way with their feet to the air sources if they were blinded by black smoke.

Two of the men in each team dragged along the

heavy rubber-and-canvas fire hoses, which quickly fattened and straightened out from the pressurized water pumped into them. The men all crouched, because in a real fire the intense heat rises and the relatively cooler area is low to the deck.

The nozzle man aimed the thick brass nozzle at the fire and simulated opening up the spray, while the hose man lifted the heavy hose onto his shoulder to help the nozzle man point it. In a real fire, the two would shift jobs every few minutes so the heat wouldn't overcome the nozzle man. To simulate having to fight the fire in a thick haze of smoke, the hose and nozzle men had light blue hospital hairnets covering their faces so they could barely make out anything ahead of them.

That was why they had the "nifti" operator directing them. The third man on the team carried the nifti, the Navy Firefighting Thermal Imager, a fat cylindrical tube with a pistol-grip handle that he held to his eyes to see through the dense smoke. The infrared camera inside the yellow cylinder gave the nifti operator looking through its eyepiece a black-and-white picture of the fire scene. Hot spots from flames showed up white on the scope.

The crew used the same procedure for fighting every fire. A rapid-response team arrived at the scene first, wearing just the EABs, the hoods, and the gloves that they had donned quickly in order to get water from the hoses pouring on the blaze as soon as possible. The sub's air banks had enough to keep the crew alive for days, breathing through the EABs. But the men wearing them had limited mobility. Every time they wanted to move a distance longer than the length of the hose, they had to disconnect it, hold their breath, then dart to another outlet to plug it in.

So within minutes, sailors in heavy yellow jumpsuits made of thick fire-retardant canvas and Gore-Tex
clomped down from both ends of the passageways to
relieve the quick-response teams. The sailors in jumpsuits could fight the blaze longer and be more mobile.
Each wore an oxygen breathing apparatus, or OBA,
which consisted of a face mask and wide visor with
rubber tubes connected to a small black canister
strapped to the chest. The canister took the carbon
dioxide the sailors exhaled and, using potassium su-
peroxide, chemically reprocessed it into oxygen,
which they could then inhale.

With a hood covering his head and an oxygen
breathing apparatus strapped to his chest, Boyd
stooped over the nozzle man for the team at the forward end of the passageway, who simulated spraying
down tubes nine and ten. Hines, another one of the
missile chiefs and now a nifti operator, bent down just
in front of Boyd, peering through his scope and pointing out the heat sources. He also had an OBA
strapped to his chest. The passageway filled with the
muffled shouts and grunts of men moving with difficulty in cramped quarters. The crewmen communicated with one another by tapping on shoulders or
yelling through their masks. Their chests heaved now
from the exertion of dragging the heavy hoses and
squatting on their haunches. The men in the bulky
fire-fighting suits were dripping wet inside. They
sucked hard on the warm oxygen generated by their
OBAs.

Hines reached down and grabbed the latch for the
floor panel in front of him and pulled up the panel. The
opening exposed a maze of gray pipes and black tubes
underneath. He ordered the nozzle man to point the

hose down into it to simulate spraying the bilge area under the deck.

Twenty minutes into the drill, Kinman and Brown decided that the teams had put out the blaze. But the nozzle and hose men, by now breathless, continued pouring water under the deck plates to cool off the still-smoldering rubber and plastic. As Dille spread out green rags to simulate wet areas from the hose water, Brown scribbled notes on a pad. Speed was the key to fighting any fire, and the hose team for the forward end of the passageway had been too slow in arriving.

Volonino, who was timing the response up in control, wouldn't like that. Practically all fires start small, so it only makes sense to extinguish them rapidly. If the crew doesn't, within minutes a blaze could spread to other levels of the sub and a skipper could have a disaster on his hands. What started out as a relatively small electrical fire aboard the Navy's USS *Bonefish* attack sub in 1988 quickly turned into a full-scale conflagration that killed three sailors and forced the crew to abandon the ship. Volonino wanted fires to be out within three minutes of their discovery. This one had taken far longer to extinguish.

With the blaze gone, the teams now searched for injured crewmen. The first priority in a submarine fire is putting out the fire to save the ship. Treating burn and smoke victims always comes second.

Hines climbed down into the opening he had made by lifting the floor panel and pointed his nifti underneath the deck. He searched for hot spots, smoldering ashes that might flare up again after the hose teams stood down—always a danger with any sub fire. Hines found one that Dille had simulated and shouted up to Boyd through his mask, "I have a hot spot."

Boyd ordered the aft hose team to push its hose under the deck to the small space between tubes nine and ten. "Did you see where the hot spot started?" he yelled through his mask. Hines could barely hear the question, but he pointed to where he thought Dille had placed the heat source. Boyd turned around and shouted the location to Chesney, who was the phone talker for this drill and was behind the executive officer.

Sweat poured down Chesney's forehead, and his muscles ached from the tension of being Boyd's phone talker. He had received a basic fire-fighting course during his sub training at Groton, but he hadn't had the advanced course at Kings Bay, so he was only marginally qualified to perform in this drill. With the blue hair net pulled over his mask, he had a helluva time finding the jack along the passageway to plug in his phone line so he could relay Boyd's reports to damage control central. Then Chesney kept cutting Boyd off in midsentence when he repeated the executive officer's reports to confirm that he understood them—which only made Boyd more irritated. Chesney now shouted the hot spot's location as best he could through his mask into his phone, hoping that damage control central on the other end of the line heard it.

Hines, by now exhausted, bent over and made another sweep with his nifti under the deck looking for hot spots. There didn't seem to be any more. He stood up again, his lower half still under the deck, and reported to Boyd that he thought the fire was entirely out. Now they had about four inches of water in the bilge.

Boyd ordered Chesney to pass along the report to DC central, along with an initial assessment of the fire's damage. If the blaze had gotten completely out of control, the sub could have surfaced and the crew could have piled out of the topside hatches as a last resort. But in this case, the *Nebraska* ascended to periscope depth and the induction mast was raised to begin sucking in fresh air and ventilating out the smoke. The induction mast was nothing more than a large pipe with a mushroom-shaped valve on top of it that allowed the sub to snorkel. Once the smoke had cleared, the rest of the crew in the missile compartment, all of whom had been wearing EABs for the emergency, could finally take off the uncomfortable masks. They could fight a war wearing the EABs, but it was bothersome having to disconnect a tube from one air plug and reconnect to another one every time they had to move.

Hines remained at the missile passageway with his nifti, looking for hot spots that might flare up as fresh air circulated back into the compartment. New teams of nozzle and hose men, wearing EABs, replaced the worn-out men in the fire-fighting suits whose portable canisters had about run out of oxygen. The hose teams stayed with Hines in case they were needed to deal with a reflash. Brown decided that wouldn't happen this time. The crew would have its hands full as it was, cleaning up—which could be a messy chore after a missile compartment fire.

The missile section was being quarantined for the moment so anyone inside it could not leave to contaminate the rest of the crew. In a worst-case fire among missiles, there is always the danger of the tubes leaking radioactive contamination from the extreme heat. Anyone inside the compartment during the casualty is considered potentially contaminated and a threat to spread it to the rest of the crew. The only exit now was at the compartment's second level on the port side. There a decontamination team had spread out a yellow plastic mat on the long passageway in front of the missile control center. On it, tired, sweaty firemen, with their bulky suits unzipped in the front to let in cool air, stood waiting to be stripped of their gear. At the forward end of the passageway was Barrass, dressed in a yellow anticontamination jumpsuit and hood with a blue doctor's mask over his face and orange rubber gloves on his hands. He helped the next man in line take off the fire-fighting equipment, which was placed in another room to be sprayed and cleaned later.

Once out of the fire-fighting suits, crewmen turned right at the end of the passageway and walked to the next station just past the administration office on the left, where another sailor in a yellow jumpsuit scanned them with a "frisker," a sort of high-tech Geiger counter. There were different types of radiation: gamma, beta, alpha, and neutron. The *Nebraska* had dozens of friskers on board that could detect each type. The crewmen decided to use the alpha instrument this time because it would be the most sensitive and would respond to the smallest amount of contamination. If the frisker found nothing, its operator lifted the maroon-and-gold rope that cordoned off the passageway and let the crewman out of the line. If radioactivity was detected, the crewman was sent to the officers' showers at the end of that passageway, which had a metal sign designating it as the decontamination station. Beside the shower hung a red decontamination bag, which looked like a long coat bag, with bandages and gauze stuffed in its many pockets. Another frisker sat underneath the bag. Most contaminants could be washed off with soap and water. But those sick or injured in the fire went to the "JO jungle."

Doc squatted on his haunches in the jungle with his own Geiger counter, which he swept over "Herbie," who was laid out on a stretcher. The JO jungle was the three-man stateroom to the right of the administrative office, reserved for the three junior officers who had been in the Navy the longest, which in the *Nebraska*'s case meant Kinman, Habermas, and Hardee. The stateroom also doubled as a sterile treatment room with an oxygen tent for burn and contamination victims. Herbie was a heavy mannequin dressed in a poopie suit, which the crew used as its all-purpose human casualty for drills. And Doc was Chief Bob Philbin, the *Nebraska*'s hospital corpsman. No one ever called him Bob, or Philbin, or Chief Philbin. Just Doc.

Doc was already sweating in his yellow anticontamination suit and mask and rubber gloves. He finished scanning Herbie with the frisker, then began quizzing him, as he would a real patient, to see if he was conscious. "Who are you?" Doc went on. "Where are you? . . . Naw, you're not in Kosovo."

The decontamination drill finished, Doc stood up, pulled back the hood from his jumpsuit, yanked off the face mask and rubber gloves, and looked down at Herbie. "Get back to work," he growled.

The sailors standing at the doorway chuckled. That was typical Doc. He had the bedside manner of a drill sergeant. When you saw Doc, he handed out one of two things for what ailed you: If it hurt above your waist, you got Motrin; if the pain was below the waist, you got foot powder.

Malingerers learned quickly that they didn't have a

gentle ear in Doc. In the month before the *Nebraska* set sail, a dozen sailors had come to his office on base with ailments they were sure would keep them from going to sea. It always happened just before a patrol. The hypochondriacs lined up at his door with invented aches and pains. He called them the "5 percent club"—the 5 percent of the crew who seemed to take 85 percent of his time. When one of these men walked in, Doc's nearly bald head, which had only a gray stubble covering it, would turn beet red. He'd squint at the man through his wire-rimmed glasses. His grizzled face would scrunch up. And he'd give them all the same prescription: "Suck it up! You're going to sea. That's what you signed the contract for."

The crew figured that all that time with the Marines had made Doc tough as leather. Philbin had joined the Navy late, at the age of twenty-six. After a year of medical training, he packed off to Camp Lejeune in North Carolina, to serve as a corpsman in a Marine battalion. (The Marines use Navy corpsmen as their combat medics.) Philbin loved being with the leathernecks those three years. They were disciplined, hardworking, no complainers among them. Several duty stations later, however, Philbin finally decided he wanted to be in the submarine world.

At first, Navy personnel bureaucrats said he was too old at thirty-five to be starting as a corpsman on a sub. So Philbin served on surface ships while he waited and enrolled in more medical schooling along the way. By 1990, he was the equivalent of physician's assistant and paramedic with training that amounted to the first two years of medical school. By 1996, the Navy was desperate for sub corpsmen, so it ignored Philbin's age and assigned him to the *Nebraska* at forty-two.

His wife, Marion, had grown used to their long-distance marriage by then. Every time he went to sea, she bought another pet. Marion now had twenty-one cats, three dogs, three fish, two tarantulas, and a turtle.

Doctoring under the sea in a Trident has its own special challenges, as Philbin found. Studies of humans in underwater environments have been carried on for more than four hundred years, but only in the last fifty, particularly with the advent of the nuclear-powered vessels, has submarine medicine become a specialty. A Trident depends on human biology as much as it does on nuclear physics.

Even though they work in steel tubes with no sunlight, surveys have found that the Navy's submariners tend to be healthier than their counterparts on surface ships. Nevertheless, sub life does produce peculiar medical problems. Carbon dioxide levels have to be carefully regulated in the vessel, and Philbin made sure all the engine room technicians wore small gauges on their belts that measured the radiation level they received. Not all sailors can adjust their body clocks from a twenty-four-hour day to the eighteen-hour cycle aboard a sub. Submariners tend to gain weight and have higher cholesterol and blood pressure levels from the sedentary lifestyle on patrol—one reason the Navy cut back on fat in the menus and put exercise machines in the Tridents. Philbin would walk around and poke bellies when he noticed girths beginning to widen. On one patrol, the officers and chiefs held "battles of the bulge" to see which group could lose the most weight.

There tend to be more cases of myopia and loss of distant visual acuity among submariners, because, being in confined quarters for long periods, they never have to focus on faraway objects. Philbin kept plenty of ointments and powders on hand to treat skin rashes, dandruff, chapped lips, and athlete's foot, all prevalent because of no sunlight and the dry air. On every patrol, Philbin also had at least one "psych case," as he called it—a sailor threatening suicide or acting screwy from being cooped up in the sub so long. The skipper transferred such men off as quickly as he could.

Philbin could prescribe drugs and perform emergency medicine. He was constantly stitching up heads gashed by hatches and hands cut up in machinery. He had a small office on the second level in the back end of the missile compartment, which contained a bed for examining patients, a sink, a computer terminal, emergency response bags, and lockers filled with drugs, bandages, syringes, and operating instruments.

But once on patrol, Philbin wasn't equipped to handle a major medical emergency and he couldn't phone a doctor for help. If he had to, there was computer software aboard with step-by-step instructions for surgery. The wardroom where officers ate their meals could be converted into a hospital operating room. But cutting someone open was the last thing Philbin would ever consider. On covert operations during the Cold War, sailors rarely were evacuated for medical emergencies. At a point of no return in a patrol they were told that if they became seriously ill, they would die on the boat. A dead sailor's body would be bagged and stuffed in the freezer until the sub returned home.

Today, the Navy gets them off the boat, but it isn't an easy operation. If a Trident has to go off alert in order to surface with a medical case, another sub has to step in to cover its target package. Transferring an ailing man from a sub to a helicopter or another ship on the high seas can be dangerous. It can take up to four days to make the arrangements. If a crewman suffers a major heart attack or stroke, he does not have a good chance of surviving.

That was why Philbin tried to make sure the crew was healthy before the *Nebraska* set sail. Fortunately, most were young and fit. But the chief nevertheless screened every sailor's medical record before leaving port to try to spot potential ailments. He paid close attention to dental records. Treating dental problems was next to impossible on patrol, and the last thing he wanted to tell Volonino was that he had to surface the sub because some idiot who hadn't gone to the dentist back in port now had a toothache. Philbin at one point walked around the sub with a large pair of greasy pliers which he threatened he would use to yank out a bad tooth. The crew half believed he would.

The yellow mat had been stowed away and the ropes that had cordoned off the decontamination area were taken down. The mannequin was dumped on one of the empty beds in the second-level ghetto for the junior seamen. Doc finished packing his decontamination bag and rolled it up tightly so it became a red duffel bag with a white cross on its outside. He handed it to a sonarman, who would store it in the sonar equipment room at the front of the sub.

Volonino came around the corner past the admin office and saw his corpsman tidying up.

"How'd it go, Doc?" the skipper said brightly. "You save Herbie?"

"Yeah," Doc answered, pushing his glasses up his nose. "Herbie's alive and kicking."

8 • Rabbit and Wolf

Scott Shafer sat at the end of his bunk facing missile tube number twelve, sewing a dolphin insignia on a blue poopie suit. He had his left knee extended, hoping that would relieve the pain. He had slipped rushing upstairs the other day and banged his knee, which had left it slightly swollen with a large black bruise. The knee felt better today but it was still tender. Shafer had been too consumed with the newness of sub life to pay it much attention. The poopie suit on which he was sewing the dolphin insignia wasn't his. It belonged to Seth Magrath, his "sea daddy." New sailors usually picked up sea daddies on the sub who looked after them and answered questions as they adjusted to life underwater. Magrath was a petty officer second class and an A-ganger. Shafer wanted to show his gratitude.

Miguel Morales, one of the new seamen, crawled out of his bunk next door and yawned. "Hey, Shafer, you eat yet?" Morales asked, stretching his arms.

"Oh yeah, chow," Shafer said, almost startled, looking at his watch. He'd been so engrossed in his sewing he had forgotten it was nearly noon that Friday. In twenty minutes he was due in the control center to start his watch. He would be a messenger running errands for the chief of the watch.

"I'll probably skip chow and go straight up to control," Shafer said after thinking about it for a moment.

Morales looked at him mystified, stretched once more, then ambled to the crew's mess in the front of the boat. Morales was new but not so green that he didn't already know that nobody missed meals on the boat. Particularly not to begin your watch early. Meals were the highlight of each day, the breaks in the dreary monotony.

His sewing finished, Shafer sprinted up the ladders to the first level at the control center. Maybe hiking around the ship would loosen up his knee.

But Shafer wouldn't be running errands for several hours. The *Nebraska* was about to go to war.

Volonino also walked briskly into control about the same time Shafer did. He reached for the loudspeaker's microphone clipped over the conn near the periscope.

"Attention," he began after keying the mike. "I've just received a top-secret message from STRATCOM. There's been a change in our mission. Two Russian submarines manned by rogue officers have come to the U.S. East Coast to threaten shipping. Our mission now is to take out both of those submarines."

Volonino clipped the mike back to its overhead cradle and bounded down from the conn. "Hah!" he said cheerily, clapping his hands.

No one else in control appeared too excited, however. They could all see Kinman with that silly red ball cap on his head, standing at the forward entrance to control, pecking away on a skinny black Panasonic laptop computer. Kinman was booting up Indy for a drill.

Sometimes another sub would come out to provide "rabbit services" so the *Nebraska* could practice tracking it for a torpedo kill. Or the *Nebraska* would be the rabbit, mimicking Russian submarine tactics, so a fast-attack sub could hunt it in torpedo war games. In those cases, the *Nebraska* intentionally made a little noise; if it operated silently, as it would in battle, the war game would be boring for the hunter sub because it would never find the Trident. Navy subs also routinely stalk merchant ships around the world so their sonarmen can practice tracking and sneaking up on a vessel. The merchants, of course, have no idea they are being hunted.

Today, however, the *Nebraska* used Indy for its torpedo war game. Indy was a computer program developed for Navy subs. It fed tonal signals to the sonar shack's headphones and video screens to simulate enemy subs lurking nearby. Typing away on his Panasonic, Kinman could create noises so it appeared to sonar that enemy vessels were maneuvering against the *Nebraska*.

Volonino's crew spent almost as much time practicing torpedo tactics as it did firing ballistic missiles. During their Navy careers, Trident submariners routinely serve tours on fast-attack subs and vice versa, so they have to be skilled in firing both missiles and torpedoes. The men of the *Nebraska,* in fact, boasted that they had no trouble holding their own in war games against the fast-attack sub hunters, who specialized in attacking vessels with torpedoes.

During the Cold War, however, a Trident would have fired its torpedoes in battle only as a last resort. If it couldn't escape and hide from a Soviet hunter, it would have launched torpedoes in self-defense. Or, after it had rocketed off all of its ballistic missiles, the Trident would become a sub hunter itself, looking for Russian ballistic missile subs to destroy with its torpedoes, if possible, before they had a chance to launch their own missiles against the United States. Otherwise, the Tridents were considered too valuable, with their twenty-four nuclear-tipped rockets, to be risked in a conventional war with other vessels at sea.

Lately, however, the *Nebraska* had been playing in more tactical exercises to test its torpedo skills. The crew suspected that with the Cold War over the Navy had softened a bit on its resistance to using the Tridents in torpedo attacks. If a rogue nation's submarine threatened the United States and a Trident was the only sub nearby, maybe it would be sent into battle. But the Navy would still think long and hard before risking a valuable Trident in an underwater brawl. The admirals shuddered at the thought of one of their ballistic missile subs being sunk or, even worse, captured by a rogue warship.

Not that that would ever happen. Just how many submarines the U.S. Navy needed now that the Soviet Union had disintegrated had become a raging debate in Washington. The admirals insisted that the some sixty they had now, half the number during the Cold War, was the rock-bottom lowest they could have and still protect the country. More than forty nations had submarines, many of them none too friendly, like Iran, China, and North Korea, and they were constantly buying newer, quieter models. Russia, the Navy brass warned, continued to sail stealthy attack subs like its sleek new Akula class of vessels, which were as silent as America's best submarines. And the Kremlin planned to build even more sophisticated boats with improved propulsion and weapons systems. All the while, the U.S. Navy's antisubmarine warfare capability had degraded, and the admirals had fretted.

Skeptics found the picture far less dire. Most countries couldn't submerge their subs for more than two days at a time, and their diesel engines were brass bands compared to America's silent fleet. Even the Russian Akulas were quieter than their U.S. counterparts only when they puttered along at slow speeds. In fast-moving combat, they made enough noise for the Navy's sonarmen to hear them. And while Russian shipbuilding continued, cash-strapped sub captains in the fleet had to find towns that would "adopt" their boats and donate bags of potatoes, cabbage, and beets so their crews could eat.

The old-timers aboard the *Nebraska* could remember the days when Soviet and American subs played rabbit and wolf. Russian subs loitering along the American coast would try to pick up the boomers sailing out of port to trail them during a patrol. The boomers always quickly lost their wolves once they sank into the water. Likewise, American fast-attack subs chased Soviet subs around the world, trying to disrupt their patrol plans by intercepting them at different points.

Occasionally, a Russian sub still went hunting, mostly in the North Atlantic or Pacific near its own borders. Naval Intelligence still tracked each one leaving port and would radio a warning if any had even a remote chance of sailing near the *Nebraska*. The Navy considered it a strategic threat to U.S. national security if a Russian wolf ever got close to one of its Tridents, even now. A pack of fast-attack subs would be rushed to the area to find the intruder and chase it away.

But the real games of rabbit and wolf were now rare. The *Nebraska* crew would grow fat and lazy if it didn't have U.S. subs to engage in mock battles. The Tridents' most skilled adversaries nowadays were other American submarines. In fact, they were practically their only adversaries. Other countries didn't even bother to hunt for the Tridents. It would be a waste of time. The U.S. boomers rarely sailed in seas crowded with foreign subs. In the lonely oceans where they patrolled, a foreign warship had little hope of finding them because they were so quiet.

Twenty minutes after Volonino made his announcement, Jason Bradfield, one of the sonar operators, hurried into the control center with the report that the towed sonar array trailing the sub had just picked up the noise of a suspicious contact. Freeland was taking his turn at the conn as the officer of the deck. Bradfield couldn't say for sure who the contact was, but he guessed it was another submarine and probably one of the Russian boats.

That was good enough for Freeland. He ordered Albea, who was his inboard planesman, to turn right fifteen degrees so the *Nebraska* for the moment wouldn't be on a collision course with this mystery vessel. Then he grabbed the overhead microphone for the sub's speaker system and announced: "Station the fire control tracking party."

The command brought out the *Nebraska*'s A-team for hunting another sub, the best sailors Volonino had on board to man each station in what would soon become an incredibly complex battle.

Harry Ganteaume was the first to bound up the stairs on the port side of the control center. He replaced Freeland as the officer of the deck. During a torpedo battle, the OOD's job becomes critically important. While Volonino maneuvered the boat for the attack, Ganteaume would make sure it didn't collide with other vessels nearby, or strike bottom, or stray into the kill zones of enemy warships. In other words, he would keep them alive, not only in the fight, but also to fight another day. Ganteaume would carry out the secret evasion tactics the *Nebraska* used after it fired torpedoes. This way the sub could hide if its quarry counterattacked. Harry had become more than a safety officer, however. During a torpedo attack, he never left Volonino's side, whispering advice into his ear on battle tactics.

Freeland hustled off the conn and forward to the sonar shack. A weapons officer normally would be punching the buttons to fire the torpedoes, but Freeland was one of the most senior of the lieutenants on board with the most training in sonar operations. For torpedo attacks, Volonino wanted him in charge of the *Nebraska*'s ears.

The control center quickly filled up with the other war fighters, who strapped sound-powered telephones around their necks, clamped headphones to their ears, and began fussing with the dials on computer screens or the charts and graphs on plotting tables. The hardest part of a torpedo attack, particularly against another sub, is simply finding the enemy in a black world where faint sounds are the only clues to where he is hiding. Two groups in control wrestled with the chore—one using computers, the other paper and pencils.

At the center's forward row of three computer consoles, Hardee sat at the far left one, designated the "wog," for weapons order generator. For the torpedo attack, Hardee became the weapons control coordinator, punching buttons on the wog, which told the torpedo's computer where to hunt for the enemy vessel.

To Hardee's right, sitting at the attack control console, was Petty Officer First Class Richard Garvin, who would actually fire the torpedo. To Garvin's right, Bob Tremayne, a lieutenant junior grade, settled himself in front of the pri-mate console. His computer took tracking information from sonar and tried to crunch out its best estimate of the enemy sub's bearing, course, speed, and range—a complex process the submariners called "target motion analysis."

Two computer consoles sat in a row behind these three men. The left computer was designated the commanding officer's display console, which Habermas sat at, also trying to come up with his own calculations for the enemy sub's bearing, course, speed, and range. To Habermas's right were two video screens stacked one on top of the other for the digital display indicator that Boyd used. The top screen showed the same picture the spectrum analyzer broadcast in the sonar shack. On the bottom screen, Boyd could tune into any of the pictures sonar received on its other stacks.

Backing up the computer operators were the pencil-pushers, whom Brady supervised at the plotting tables. Dave Bush, a lieutenant junior grade who worked in the engineering department, stood over the geographic plot, which tracked the *Nebraska*'s position and those of other contacts. Next to him on the same table, Rob Hill, another lieutenant junior grade in the engineering department, manned the time-bearing plot. Hill marked on his chart the position sonar believed an enemy sub was in relation to the *Nebraska*—that is, the other sub's bearing—and recorded the changes in that bearing over time. (The enemy sub's course, another calculation the crew made, was the direction in which it was actually moving.)

At another table nearby, Lieutenant Joe Davis plotted the distance between the *Nebraska* and the enemy sub—that is, the range. Davis also tracked the changes in that range over time. The *Nebraska*'s supply officer, Davis found the time-range plot a welcome diversion from his day job of tracking produce and toilet paper.

Finally, there was Thorson, who stood near the forward entrance to the control center with a fistful of colored pens in front of a large sheet of graph paper that rolled down vertically from a stand. He was in charge of the time-frequency plot. This was the graph that the *Nebraska* used to keep track of all the frequencies its sonar detected from an enemy sub.

The three time plots proved extremely useful for zeroing in on the enemy. Record enough changes in the contact's bearing, range, and frequency—particularly as they relate to your own sub's position—and eventually with a little high school geometry and trigonometry you can come up with a good guess as to where the enemy is. Back at submarine school in Groton, the young officers had spent weeks in the "plotting palace" practicing how to find contacts on their charts.

Boyd had barely fitted on his headphones when reports from sonar, the plotters, and the computer operators began pouring into the earpieces. He was now the fire control coordinator. Actually, he was a combination traffic cop and final arbiter in the torpedo war. A committee, in effect, hunted for the enemy sub—a committee in which each member worked independently and each developed strong opinions about where the prey was. Freeland in sonar, Tremayne at the primate, Habermas at the commanding officer's display console, Brady hovering over the plots—each of them was coming up with his best guess on four critical pieces of information about the sub: its bearing, course, speed, and range. Pinpoint them and you had your target in the crosshairs.

Boyd's job was to pass clues back and forth among sonar, the computer operators, and the plotters, to sort through the blizzard of information and guesses he got over his headphones on the sub's position, to decide whose analysis was accurate and whose was faulty, to constantly refine the estimates, and finally to come up with the magic answer. When the "firing solution," as the submariners called it—the bearing, the course, the speed, and the range of the contact— was fed into the brain of the torpedo, it could wind its way through the dark ocean and nail the enemy sub. It could be a maddeningly complicated exercise, particularly when sonar was tracking several hostile war vessels. During a torpedo battle, Boyd darted around the control center like a pinball.

Freeland's voice now came crackling into his earpiece. "Coordinator, sonar, gain, new, towed array contact. Designated sierra nineteen, sierra twenty." Translated: there might be one sub out there, but Freeland didn't know exactly where it was.

The problem was "bearing ambiguity." The small hydrophones on the towed sonar array, which trailed out about a mile behind the sub, listened in a 360-degree circle. But the hydrophones couldn't tell whether the sound was coming from the right or left side of the array. For the moment, Freeland designated the contact sonar had picked up as two, sierra nineteen and twenty, because at this point he didn't know which side it was on.

Volonino had already walked into the control center and was setting up the maneuvering board on a ledge

behind the periscopes when Boyd relayed to him
sonar's report on the contact and the bearing ambiguity. The plastic-covered board was poster-sized, with a
large circle on it marked off by degrees with lines and
points for plotting. During the hunt, Volonino drew arrows on it with a grease pencil for where his sub and
the enemy sub were heading to help him think through
the tactics he needed to make the kill.

After listening to Boyd, Volonino ordered Albea to
put the sub into another right turn. The easiest way to
resolve the bearing ambiguity was to change the *Nebraska*'s direction, let the towed array line straighten
out again, then have the hydrophones listen once
more. Use the new readings and the old one and with
more math you could pinpoint the real bearing for the
contact.

"Attention in the attack center," Volonino said
sharply. Murmured conversations stopped. All heads
turned toward him. "We've gained two contacts: sierra
nineteen and sierra twenty. It could be a merchant or an
enemy sub. My objective is to identify this vessel. If
it's the enemy, we go to battle stations."

But Volonino first wanted to make damn sure it was
the enemy. This wasn't World War I or World War II.
Unrestricted submarine warfare was history and probably would remain history. There were no free-fire
zones out there. In today's wars, merchant ships from
countries not involved in the fight would likely be sailing among the enemy vessels. Friendly navies would
be mixed with unfriendly ones. Volonino would likely
be summoned for a surgical strike, against a single
vessel—perhaps to demonstrate U.S. resolve—then
he'd probably be ordered to back off, allow the crisis to
de-escalate, and hope that the war didn't widen. Mis-

taking a Japanese tanker for a Russian destroyer and sinking it wouldn't be too smart.

As sonar scrambled to get a better fix on the contact's bearing and identify it, Volonino began turning over in his own mind the attack strategy he would use. If the enemy was a surface ship, the battle could move quickly. Volonino would have the added advantage of being able to peek at his prey through the periscope so he could identify it and estimate with his own eyeballs its bearing, course, speed, and range. But sub-on-sub engagements could take hours, sometimes even days, as two nimble and silent adversaries maneuvered in the dark against one another in what could become a slow but deadly dance.

Kinman quietly typed away on his laptop, sending more tonal signals into the sonar shack, which by now looked like an organized madhouse. The entire sonar division, almost a dozen men, now jammed into the small, darkened room, which became stuffy and hot from the glowing video screens and so many human bodies jostling one another and talking at once. Barrass sat in front of the spectrum analyzer on the far left, his eyes straining to spot any clues on the large screen that would tell him what kind of vessel they had out there. Wilhoite now sat at the narrowband stack and Ben Dykes, a petty officer third class, operated the broadband console. Other sailors punched buttons on other computers, or stood in front of plotting or maneuvering boards marking them, or spun small rangefinder wheels in their hands, all of them trying to come up with their own best estimate of the enemy sub's position and shouting it to Snyder. The sonar chief was now the shack's supervisor. Freeland stood beside him and behind the men hunched over their screens with a phone set strapped around his neck so he could relay sonar's reports to Boyd.

The chatter in the shack never stopped. Snyder, always fidgeting, pleaded and prodded and praised and snapped at his men to pry more information out of their machines. Their moods soared up and down, from joking and burping one moment to testy exchanges the next. Snyder wanted it that way. He wanted his men debating who had the best fix on the enemy sub's position, arguing over the best tactics for the kill. The more they argued, the more they refined the firing solution.

Back in the control center, Volonino, Boyd, and Ganteaume crowded around the digital display indicator for a look themselves at the sonar screens. In low voices they began debating who had the best fix on the enemy sub's position. Chesney had been assigned to operate the TAC-4 console to their left, which displayed on a multicolored screen the direction the contacts were supposed to be moving. Chesney was supposed to phone Boyd with his guess on the contact's location, but the executive officer heard nothing from him.

"TAC-4, are you on the wrong circuit?" Boyd snapped. "I don't hear you."

Chesney was in fact speaking into the wrong circuit, which was just as well at this point, because the old bearing ambiguity still had him confused about the contact's location. Volonino came over and gently explained how he should reenter bearing data in the TAC-4 to get back into the game. "Now listen to what the XO is saying on his circuit," the skipper said calmly.

The pace picked up in control. Volonino turned to Ganteaume. "Officer of the deck, man battle stations," he ordered. Ganteaume repeated the command over the loudspeaker system.

Boyd stepped from the digital display indicator to the plot tables to Tremayne at the pri-mate console to Habermas at the commanding officer's display console. More information came from sonar, which, after the sub had made its last right turn, had resolved the bearing ambiguity.

The contact was redesignated master-one so no one would be confused about which one they were now tracking. Master-one was somewhere northwest of the *Nebraska*. Sonar, the computer operators, and the plot men hustled to come up with firing solutions. Garvin hopped out of his seat and visited Habermas for the latest he had from the commanding officer's display console. It might help Hardee with the presets that would eventually have to be fed to the torpedo's computer. Ganteaume took readings on the depth of the ocean under the sub in case they had to dive deeper to evade, then began conferring with Volonino quietly on what he thought the firing solution might be.

Tremayne and Habermas and Snyder, who had come from sonar, began handing Boyd white slips of paper. Scribbled on each one was their best guesses on the contact's bearing, course, speed, and range—with plus-or-minus margins of error. Boyd took the chits and walked over to Brady to have them drawn on the geographic plot. Boyd used Brady and the geo-plot to help him decide which chit's information, or combination of information, gave him the best firing solution. When Bush finished marking the plot, Boyd took a final look and scribbled on another slip of white paper.

"Here's my best solution," Boyd said, walking up to

Volonino. The Russian sub was more than three miles away on a bearing of 310 degrees. "Course zero-four-zero," he continued. "Speed seven knots. Sonar concurs, it's a sub."

Boyd recommended that the *Nebraska* make another turn to get into a "preferred firing position" to take the shot.

Volonino quietly quizzed his executive officer on his firing solution. Finally, he was satisfied that Boyd had the best one.

Three levels below, the torpedomen began making preparations for the shoot. Boyd turned to Tremayne, who had come up to the conn to listen in.

"Enter system," the executive officer said, handing him the fire solution. Tremayne punched it into his primate console. Its fire control system then calculated the best course for the torpedo to intercept the sub. Hardee quickly began punching the presets for that course into the wog, which fed them into the torpedo's computer.

"Captain, recommend making tubes two and one ready in all respects," Hardee said to Volonino. The *Nebraska* had four tubes at its bow for shooting torpedoes. Hardee now had to busy himself with preparations so two of them could launch.

"Make tubes two and one ready in all respects," Volonino commanded. "Helm left, fifteen degrees rudder." Albea repeated the order and turned his wheel to the left.

"Captain has the conn," Ganteaume now announced. Volonino would personally drive the sub to the point in the ocean where he thought he could take his best shot. Ganteaume would be in charge of the rest of the boat while he maneuvered. "Helm, steady course two-nine-zero."

Albea repeated the order and turned his wheel to send the sub on a course of 290 degrees.

Boyd pressed one side of the headphone to his ear, then walked up to Volonino. He had just received a report from Freeland. Sonar now was sure the contact was a Russian sub. But it was one of the older models.

That threw a slight complication into Volonino's attack plans. In the Indy scenario, Navy Intelligence had told him that one of the two Russian subs was an older vessel. But the other sub was a much more modern Akula, a far more valuable prize if he could bag it. Volonino decided to follow the older sub for the moment, but hold off killing it until perhaps he could locate the Akula.

He asked Boyd for an update on his firing solution. Master-one's range, course and speed remained the same, but Boyd believed his bearing had changed to 325 degrees.

"Attention in the attack center for a tactical brief," Volonino said crisply. "We have positively identified our enemy." But it was an older Russian sub, he explained. "Intel says there's another more valuable Akula out there. So my intention is to follow behind this contact. We'll get deep into the target's baffles. If we have to, we'll engage him with a Mark-48 torpedo." But for the moment, the *Nebraska* would follow and wait.

Volonino decided to have Hardee load torpedoes in the two tubes.

Boyd paid another visit to the plot tables. Master-one still seemed to be headed in a northerly direction.

He walked across the conn to the port side of the control center, where Petty Officer First Class Rodney Mackey stood making marks on the vertical contact evaluation plot. An African-American, Mackey also conducted the Protestant religious services aboard the sub every Sunday. But now he was listening for bearing and time information from his headphones and with colored pens writing on the plot everything the control center knew about the contacts out there from minute to minute. It helped both Boyd and Ganteaume keep a mental picture of where everyone was in the water.

Hardee reported to Ganteaume that both torpedo tubes were now ready for launch. Boyd crossed the conn again and returned to Brady at the plot tables. It looked like they had a "target zig," Brady told him. The Russian sub had suddenly changed course. It seemed to have banked sharply to the right. That didn't particularly surprise Brady. The Navy had compiled volumes of top-secret intelligence on Russian tactics. Russian sub commanders were disciplined. They were as careful as American skippers to keep unnecessary noise down in their boats. But they didn't think as three-dimensionally underwater as American captains did. And though they could always pull a few surprises, Russian submariners tended to be predictable in their movements and maneuvers, so U.S. skippers could often anticipate their zigs.

But Boyd had to again refine his firing solution. He asked Brady, Tremayne, Habermas, and the sonar shack to come up with new numbers. Then he turned to Chesney. "TAC-4, you have a solution for me?" he asked tersely.

"I'm working on it, sir," Chesney replied meekly, although he had no hope of calculating one by the time Boyd needed it.

Boyd shoved a white slip under the ensign's nose with blanks left for the bearing, range, course, and speed. "Here," he said brusquely. "So you'll know what to report the next time."

After checking the plots once more, Boyd handed Volonino another slip with the updated firing solution.

Volonino reviewed the chit. It was pretty much on target.

"Attention in the attack center," he said, looking up. "We are starting to develop a better firing solution." The Russian sub had sailed a mile and a half farther away from the *Nebraska* on a bearing of 340 degrees.

"We'll turn to that bearing more," Volonino decided. "I want to get closer to him before shooting. Then we'll engage him with tube two and use tube one as a backup." After the shot, the *Nebraska* will begin its evasion plan and monitor for incoming fire from the Russian sub. If it did manage to get off a shot before the *Nebraska*'s torpedo struck, Volonino said, "We'll conduct a full evasion. Carry on."

"Captain, tube one and tube two have been made ready," Hardee announced.

"Very well," Volonino responded.

Brady announced that the sub had zigged once more. Boyd ordered another round of firing solutions.

Volonino squatted down and sat on the edge of the platform for the conn, in front of the number one periscope, his arms folded over his knees, as if he were resting on a street curb on a lazy Sunday afternoon. But one question kept running through his mind. Where was that Akula?

Ganteaume continued receiving soundings from the

fathometer operator on the depth of the ocean in case they had to dive farther to escape enemy counterfire. Boyd kept refining his firing solution for the old Russian sub the *Nebraska* was tracking. Volonino stood and began pacing around the conn like a nervous cat. His sub was now sailing almost parallel to the course of the Russian boat.

Then it happened. Kinman typed in a new command for the Indy program. Several seconds later, the chatter in Boyd's earpiece was interrupted by a report from Freeland.

The second prey had appeared. Dykes, who had been staring at the broadband stack's screens, suddenly spotted a second thick white line forming on his green waterfall. It was the Akula. The newer sub had been creeping along slow and silent near its older brother. The *Nebraska*'s sonar had apparently been unable to detect it until the Akula speeded up and made more noise. The shack had found the prize.

"Attention in the attack center," Volonino said, now with a little more excitement in his voice. "We have a second contact, this one more valuable." The Akula he had been expecting had finally appeared on sonar's screens. Master-two was now the target of interest.

"Helm, left fifteen degrees rudder, steady course two-five-five," Volonino ordered. They would ignore the older sub and chase the Akula.

The second Russian sub also had no idea at the moment that it was being stalked by a Trident. The U.S. Navy's submarines still had an acoustical advantage over even the newest Russian models. The Akula didn't give off much sound, but the *Nebraska* was still quieter. Its sonar also was more sensitive and could detect the Russian sub before the Akula's sonar spotted the Trident. As long as the *Nebraska*'s sailors didn't do anything careless—like dropping a wrench on a metal deck, which the Russian sub's sonar would easily hear—they could safely trail the Akula from a distance.

But the *Nebraska* had to watch out that it didn't creep too close to the Akula or it would be detected. What's more, Volonino didn't want to fire his torpedo at too close a range or the Akula's sonar could spot where the bullet had come from and fire back at its source. The Trident had computers on board that used complicated math formulas to calculate the ranges at which enemy vessels could detect it. The sub that shoots first in an engagement usually wins. But if Volonino wasn't careful, his own sub could be blown to bits after the kill—"not a particularly good exchange ratio," as he liked to put it antiseptically.

Distance was now on the commander's mind, however. The Akula had popped up on the *Nebraska*'s sonar at a close distance, and the range between the two subs was closing. Volonino turned around and began drawing arrows on the maneuvering board behind the periscopes. Ganteaume stood behind him with his rangefinder wheel making his own calculations.

Volonino began firming up in his own mind his attack strategy, plus his postlaunch strategy. He had no problem now putting together a plan. Stacked up, the number of secret volumes the Navy had published on submarine tactics was taller than he was. Volonino had spent twenty years studying them and honing his battle skills in attack trainers and in exercises at sea, where practice torpedoes with their warheads removed were actually fired at moving targets. The fact that the U.S. Navy routinely shoots off exercise torpedoes is what distinguishes it as a world-class submarine power.

Most Third World countries that own subs rarely practice-fire their weapons.

The time-outs Volonino had been calling for tactical briefings became even more important now. As the *Nebraska* neared the point where it would launch a torpedo, the tempo would pick up even more in the control center. Many things would happen simultaneously. The officers and crewmen would have to act quickly on their own without specific orders from him.

"Attention in the attack center," Volonino announced again. The Akula was closing its range with the *Nebraska* and could be within several miles. The *Nebraska* had to worry now about "counterdetection." If that happened, he intended to launch immediately to surprise the Russian sub, then quickly begin the torpedo evasion.

"Helm, right fifteen degrees rudder," Volonino ordered. "Steady course zero-zero-zero." They would sail north to reach the point where he wanted to fire.

Boyd continued scurrying about, trying to have a new firing solution for the Akula quickly calculated. Several minutes later, he had one. The *Nebraska* also neared the point in the ocean where Volonino wanted to take his shot.

"Captain, I have a firing-quality solution," Boyd finally said to Volonino and handed him another white slip of paper with the Akula's bearing, range, course, and speed.

Volonino reviewed the chit, turned around briefly to check his maneuvering board, then approved the solution. Boyd rushed to Tremayne and Hardee to have them enter it into the computers and feed the torpedoes the presets.

"Firing point procedures, master-two," Volonino an-

nounced. "Tube two primary, tube one backup." One
torpedo would be shot out of tube two. If it malfunc-
tioned, a second torpedo would be fired out of tube one.

"Firing point procedures, master-two," Boyd repeated to acknowledge the order. "Tube two primary,
tube one backup."

The sonar operators in the shack began realigning
their computers so they could track the torpedo once it
was fired. Bush at the geo-plot quickly took out his
Mark-48 template and laid it on his chart so he could
follow the torpedo's path and evaluate the shot.

"Plot ready," Brady announced.

"Solution ready," Boyd followed.

"Ship ready," Ganteaume said quietly.

"Weapon ready," Hardee said.

But Volonino wasn't ready just yet. The *Nebraska*
was uncomfortably close to the Akula.

"We'll try to let the range open up a bit to enhance
our survivability," Volonino decided.

But the news from sonar wasn't good. The Akula
had crept nearer. If the *Nebraska* fired a torpedo now it
might not escape the Akula's counterfire, Boyd reported.

"Too close," Volonino agreed.

Sonar came back with an update a minute later.

"Sir, I believe we're opening the range a little,"
Boyd said, after talking on his phone to Freeland.

That was encouraging. But not for long. Several
minutes later, Boyd received bad news from sonar.

"Sir, we've lost master-two," Boyd told Volonino.
Sonar was scrambling to put another tracker on the
Akula. The sonar trackers normally followed the contacts automatically, but the Akula was so quiet the
trackers couldn't distinguish it from background noise,

and the computerized pinpointing system had been wandering.

"Resolve," Volonino said curtly. "Manually buzz bearings and obtain a new firing solution." The sonarman could bypass the automated trackers and manually send the target data to the control center, a process that was nicknamed "buzzing." Since the sonarmen now knew what the Akula looked like on the screen in their stack, they could track it with less of a noise signal than the sonar computer could.

Sonar picked up the Akula once again and Boyd scrambled for a new firing solution, but it took several more minutes.

"Attention in the attack center," Volonino announced. He would shoot when there was a little more distance between the *Nebraska* and the Akula. "Then we will conduct an immediate torpedo evasion. Carry on!"

Volonino's tactics at this point were simple. He wanted to fire his torpedo from a point in the ocean where the Akula had the least chance of hearing it until too late. Volonino therefore had positioned the *Nebraska* to trail behind the Akula; the Russian vessel would have more difficulty listening in its rear area because its own engine noise would mask sounds.

He also planned to fire his torpedo from a tube on the side opposite where the Akula was sailing. That way, the *Nebraska*'s bow would mask the noise from the torpedo shooting out of the tube. Hardee had also tweaked the presets fed into the torpedo's computer so the weapon wouldn't begin to send out an active sonar signal to finally home in on the Akula until it was close to the Russian sub. The Akula would hear the pinging when the torpedo activated its sonar, but by then the Russian boat would have no time to evade.

Finally, enough miles were between the *Nebraska* and the Akula that Volonino felt safe in attacking. Boyd had what he thought would be the best firing solution.

"Shoot on generated bearings," Volonino ordered.

Garvin, sitting next to Hardee, quickly punched buttons on his console to fire a torpedo out of tube two.

Whoosh! The crewmen in control could hear the sound in the torpedo room three floors below. In this case, the torpedomen had fired off a shaft of water out of the tube, called a "water slug," to simulate the launch of a weapon. If it had been a real torpedo, it would have been shot with a wire attached that could unravel out for as far as ten miles. Hardee and Garvin in control could send signals over the wire to instruct the torpedo to change direction if the target suddenly zigged. The *Nebraska* didn't have to fire the torpedo on a direct line to the target. The propeller-driven Mark-48 could turn corners so it could take a circuitous route to its victim. If the wire broke, the torpedo would head to the last point its computer had received from the sub.

If the target had been a ship, the crew would have armed the torpedo to have it explode under the hull to do more damage. The blast would create a whipping effect, much like what happens when a building implodes, which could break a ship in half. For an enemy sub, the torpedo was armed to punch a hole in its hull, which would be enough to destroy the vessel underwater.

Immediately after the launch, sonar began tracking the torpedo to make sure it stayed on course for its intercept point. Ganteaume put the *Nebraska* into its torpedo-evasion maneuvers. The sub could make erratic turns, dive, or speed off to escape what would soon be a very angry Akula captain. Ganteaume also had countermeasure devices he could fire from other tubes on the *Nebraska* that would spoof torpedoes the Russian sub shot.

In this case, there was no counterfire. "Sonar hears an explosion, sir," Boyd finally reported.

"Very well, XO," Volonino said and halted the drill. "Lieutenant Kinman will now read the target's actual position so you can compare it to your firing solution."

Kinman peered down at his laptop. The Akula had been more than two miles away, bearing one-one-zero, he reported. "Course one-zero-zero. Speed six knots."

Boyd looked at his last white slip and smiled. The solution had been close enough that they would have killed the sub.

9 • The Wardroom

Frank Levering stood in the third-level passageway just aft of the chiefs' quarters and wardroom. On the other side of the passageway was the galley. Large cans of tomato paste, sacks of flour, plastic bags of chicken patties frozen rock-hard, condiment boxes, and assorted vegetable crates were stacked around him as he checked off items on a form for the "breakout." Levering was the *Nebraska*'s "Jack of the Dust." In the days of wooden sailing ships, food was stored below in dirty holds and the seaman who hauled supplies out of it to the cooks was called Jack of the Dust because he always emerged from the storage hold filthy. The job title stuck, and the Jack of the Dust in modern Tridents, like a grocery store clerk, parcels out every morning the food and supplies the galley used for the next day's meals.

It was no longer dusty work. Across from the galley, a shiny clean chill box the size of a large bathroom, and packed to the ceiling, kept refrigerated items like fresh vegetables, fruits, eggs, and milk. Next to it was a freezer four times as large, also packed to the brim. Another nearby space the size of three master bedrooms contained seventeen large food modules filled with dry goods. A crane had lowered the modules through hatches into the sub when it was in port.

Food and morale are what limit how long a Trident can remain underwater. For this patrol, 119,000 pounds of food had been loaded into the sub, which would allow it to remain submerged for 132 days if necessary. Everything came in bulk: 3,700 pounds of baking potatoes, seven large cases of carrots, 500 pounds of onions, 900 eggs (each thinly coated with wax to keep them fresher). Totaled, it cost about $150,000, or $5.81 per man per day.

Levering found the job tolerable enough. Now twenty-one, he had joined the Navy about three years ago when it became clear to him that working as a part-time meter maid after high school wouldn't be enough to pay for college. On the *Nebraska,* he found that the work hours were long, twelve to eighteen a day, and serving the officers their meals in the wardroom, his other chore aboard the sub, was certainly not rewarding. But a Jack of the Dust was considered as important as a missileman. Levering had had to have a letter of appointment from Volonino. A skipper always screened his Jack of the Dust to make sure he didn't skim off the top; one meal of crab legs cost $1,000. And the Navy had offered him $30,000 to reenlist—an offer he planned to take up so he could pay for school and study to be a chef.

Levering double-checked his form with the recipe card from the galley, which listed the ingredients needed for the next day's meals. He adjusted his gold-wire-rimmed glasses on his nose. He wore a white paper server's hat and a blue polo shirt with the *Nebraska* seal sewn on it, which sagged some over his heavy frame. The galley personnel all wore the polo shirts instead of poopie suits.

All the items were accounted for, so he began loading the dry goods into a small locker next to the freezer and the chilled items into a refrigerator in the galley.

The noon meal was already prepared. Jason Duff had awakened at three-thirty Friday morning to begin cooking it. Now he shoved trays of dough into the oven for the 450 rolls he'd need for dinner. The baked ones he pulled out smelled and tasted heavenly because he always dabbed garlic butter on their tops before they were cooked.

Duff had a mustache, and splotches of flour decorated his blue polo shirt, which stretched tightly over his big belly. A month earlier, he had been aboard the USS *Narwhal*, a fast-attack sub based in Norfolk, Virginia. The chief of the boat, Dave Weller, had managed to talk Navy personnel detailers into having him transferred to the *Nebraska,* which needed another cook. The Navy was short of cooks, and Duff had a reputation in the fleet for being a good one, which made him an even more valuable find. Duff had cooked for five different boats. He knew a lot of the *Nebraska* sailors from tours he'd done with them on other subs. Many of the sailors here had tasted his cooking before and liked it. Off duty, he worked part-time at a resort restaurant on the Florida coast, which helped him discover new recipes to try out on his crew.

Duff seemed perpetually chipper at sea. He dished out generous portions and liked to banter with the sailors in the serving line. He joked with the captain about fattening him up. Duff considered himself one of the sub's morale officers, which wasn't far from the truth. Nothing could sink the mood of a boat quicker than lousy meals.

Fortunately for the *Nebraska*, the food the Navy serves on subs tends to be a cut above what surface sailors eat, or so submariners think. Steak, crab, and lobster are occasionally on the menu. Trident sailors are considered even pickier about what they eat than their brethren in the fast-attack subs. When you're cooped up for months, the four meals served every day—breakfast, lunch, dinner, and midnight rations— become a respite, a sanctuary, a break from the monotony. The crew's mess becomes the neutral territory where nukes and conners can escape and mingle. The *Nebraska* cooks were under orders to wear smiles when they ladled out the chow—and they usually did—because theirs might be the only friendly faces a seaman saw all day.

Duff considered the *Nebraska*'s galley a palace compared to the cramped kitchens he'd worked in on fast-attack subs. It had two ovens, a deep-fat fryer, a grill, three ten-gallon steam-jacketed kettles, a food mixer, two microwaves, and a steam line for keeping food warm when serving. A forward pantry used for serving the wardroom also had a grill. Aft of the galley was the scullery, where the sailors who were stuck with cranking hand-washed dishes, then sent them through a heated sanitizer to kill bacteria.

Time and how the sub operated could make cooking a challenge. Soup and sandwiches sometimes had to be served when the *Nebraska* sailed ultra-quiet and on reduced electrical power. After about three weeks, lettuce, fresh vegetables, and fruits spoiled, so the cooks had to substitute with canned goods. Powdered milk almost always had to replace fresh milk. Occasionally the cooks ran out of a particular canned vegetable or some other item, but Davis, the supply officer, tried to avoid that embarrassment, particularly for the three considered critical: salt, flour, and coffee. If it happened to surface and meet another Navy vessel at sea—which it had been doing more often in recent years—the *Nebraska* would arrange for a "boat pack" to be delivered with twenty-four gallons of milk, loaves of fresh bread, and several cases of eggs, fruits, and vegetables.

It was impossible to satisfy 162 different tastes, but the cooks tried. The sub had a menu review board made up of officers and enlisted men. The cooks were told to ask sailors what they thought of the meals, and the sailors were never shy about griping. Two entrées, rolls, soup, and salad (when it was available) were always served for lunch and dinner. At breakfast, sailors could have eggs cooked to order, omelets, bacon, hash browns, cereals, and pastries. The most popular foods were fried chicken, hamburgers, hoagies, and pizza. The cooks believed the sailors would eat these four every day if the Navy dietitians, who had to approve the menus, would let them. Liver was unpopular. Chinese beef and broccoli bombed.

Despite the *Nebraska* galley's best efforts, the crew always tired of institutional food as a patrol dragged on. Canned vegetables got old, and the combinations (egg drop soup and sloppy joes for dinner) could get weird. Meals quickly earned earthy nicknames. The sailors called roast beef "baboon ass" and Salisbury steak "trail markers." A fried fish patty was called a "three by five," a chicken patty was called a "hockey puck," and a veal patty was a "vent cover." Hash browns in the morning were "toenails," and the sausage links next to them were "cat doo."

Though the galley had a drink dispenser, sailors kept cases of sodas in their racks just so they could enjoy the delight of popping open a personal can. Saturday nights broke the monotony when the cooks took off and one of the divisions volunteered to bake hand-tossed pizzas by the dozen. On other special nights, the chiefs might take over and cook the prime rib—about the only time you could get a rare piece of beef on the boat, the sailors grumbled.

Levering padlocked the chill box and freezer so no one could filch food from them, then crossed the passage-way to the wardroom pantry so he could get ready to serve the officers their noon meal. The sailors never took their gripes with the cooks too far, he thought. They knew there were three people on a sub they didn't want to piss off: the yeoman who processed their paychecks, the corpsman who took care of them when they were sick, and the cook who fed them.

The officers had finished their salads in silence, and Levering began removing their plates. Normally the junior officers would be more chatty in the wardroom during lunch, but the squadron riders on board were eating meals with them and the *Nebraska* officers didn't know them well enough to let their hair down. Best to speak only when spoken to while the bosses snooped around.

The wardroom was the size of a child's bedroom, but not particularly cramped. The ten-foot table stretching down its middle was now covered with a gold cloth, the ship's silverware and china, plastic pitchers of sodas, condiment bowls, and baskets of hot buns. Each officer kept his cloth napkin tightly rolled in a pewter ring, and the napkins were all lined up on a cabinet on one side of the table to be grabbed as the diners walked in.

There were ten chairs, with light brown vinyl covers over them, arranged on the sides of the table. At its far end, there was a couch along the forward bulkhead where other officers could crowd in. Volonino always sat at the head of the table, with the galley and wardroom pantry behind him. A phone also hung on the wall behind him that he could easily reach for calls from the control center, which constantly interrupted his meals. Hanging from the ceiling to his right was a small gray box with red digital displays indicating the sub's speed, depth, and course.

The wardroom was nicely decorated. Fake wood paneling covered the bulkheads, cabinets, and drawers. A glass cabinet to Volonino's right displayed a large oval silver serving tray donated by the state of Nebraska, along with a silver tea set that had belonged to the original *Nebraska* battleship. Below them was a glass-encased box with three footballs from the University of Nebraska that had been in three national championship games the school had won. On the wall behind Volonino hung plaques presented to the sub by foreign navies. On the wall to his left was a color print of the *Nebraska* firing one of its missiles and a cabinet with a coffee machine, television, and VCR. Down the bulkhead on Volonino's left hung a University of Nebraska pennant, a ship's clock, and a brass plaque commemorating the sub's first officers. On the forward bulkhead facing him hung an oil painting of the *Ne-*

braska, a watercolor of an eagle with the American flag in the background, and a framed copy of a letter John Paul Jones wrote on the topic "Qualifications of the Naval Officer."

Levering brought out trays and bowls heaped with turkey and noodles, honey-baked ham, sweet potatoes, beans, and steamed corn. Volonino helped himself first, then passed each bowl and tray on. Dining was as formal aboard the *Nebraska* as it has been on the Navy's surface ships for centuries. Etiquette was strictly observed. Officers arriving late asked Volonino's permission to be seated, and if they had to leave early, they asked to be excused. Normally Boyd sat at the first seat to Volonino's right, but since Captain Hunnicutt was aboard, that place was reserved for him. Boyd tucked into the second seat on the right.

Meals weren't the times to probe the officers for their thoughts on nuclear war or the end of the Cold War or the ultimate mission they were assigned. That would violate the wardroom three don'ts: don't talk shop, don't talk politics, and don't talk religion.

The first don't got violated all the time. The officers brought their work to the table constantly. They joked about jobs on the sub, usually in the military acronyms of their rules and regulations; civilians would have no earthly idea what was funny. Occasionally politics surfaced in discussions. Habermas usually took the conservative side, especially on the role of government in society. Kinman enjoyed being provocative just to get under Volonino's skin. More frequently, the weighty subjects they argued over were economics, mutual funds, whether television was the downfall of America's youth (one of their most heated discussions during the last patrol). Religion was never brought up, and no one could ever recall a philosophical talk at the table on nuclear war. If you wanted to ruminate about blowing up the world, do it on your own time. The officers considered the meals just as much a break in their routine as the sailors did. Nuclear war wasn't necessarily a taboo subject. But it was probably the surest way to kill a wardroom conversation if raised. So I kept silent and mostly listened.

Occasionally, though, I found openings to bring up some controversial subjects during meals. Before the *Nebraska* left port, I had sat in the wardroom eating chicken fingers for dinner with Kinman and Habermas. Though the vessel was still tied to the dock, the two lieutenants had the duty that night, so they were stuck on the sub until the next morning, supervising the skeletal crew aboard. They had been telling me Naval Academy war stories, one about how a midshipman had tried to embarrass a Secretary of Defense, who was addressing an Annapolis assembly, with a provocative question about gays in the military.

"What a dumbass," Habermas had said. He stood up from the table and headed for the wardroom entrance. He had rounds to make in the engine room.

When Habermas closed the door behind him, I decided to try out a question with Kinman, who had stayed behind. "What do you think about gays and women serving aboard subs?" I asked, flipping open a notebook as unobtrusively as I could.

The other crewmen weren't shy with their answers to the same question. No one minded closet gays serving on board. But if a sailor was openly homosexual, it wouldn't work, most all insisted. Gays were incompatible with military service, they believed, particularly in this confined space. It would just be one more headache they didn't need, the officers said.

Women wouldn't receive a much better reception. The problem was logistics, the crewmen maintained. Tridents simply were not equipped to handle females. The sub's sailors had just two heads with a total of four showers and six toilets. That was for 130 men. Give the few women who would likely be on board just one of those bathrooms and the number of places the other hundred-odd men had to pee and wash was cut in half.

Women would have to have separate sleeping quarters. But they would also have to be assimilated into all echelons of the sub. Logistically, those two requirements couldn't be accommodated, the submariners argued, unless all the women were put in the chiefs' quarters. But the chiefs would mutiny if they had to go back to the crew's bunk rooms and give up their choice space to the junior women.

Gays aboard subs? The men shuddered at the thought. Women and men stuck underwater in a Trident for three months with no privacy for anyone? Not a good idea. Besides, what wife in her right mind would send her husband to sea with women aboard the sub?

Kinman thought about the question for a few moments, toying with the food left on his plate.

"I don't follow their argument," he finally said. "The same argument was made in the fifties for why you couldn't have blacks aboard. They were saying you couldn't have blacks and whites working together in the galleys or things like blacks can't give whites orders."

Logistics prevent it? "Bullshit!" he said, looking at me intently. "We're smart guys. This is the greatest navy in the world. If it were a priority to have women on board, we could build another head in the missile compartment. Or you could designate times that females could use the existing showers. Right now, the sub force is dying for enlisted people and officers and here we are excluding fifty percent of the population. That's just fucked up!

"Oh, they say the women can't turn the heavy valves on board. Well, we've got men too small to turn the heavy valves, but we compensate for the weak males and we don't say a word about it. The reason we don't have women in submarines is because we don't want them. It's just cultural resistance. It's just resistance to change."

Kinman was now wound up. He had changed his views on gays and women in the military 180 degrees from his days at the Naval Academy, when he followed the crowd. He might have been born in Athens, Georgia, but he was no cracker. He had come from an integrated high school. Blacks and Latinos had been his friends growing up.

"The same rules should apply to gays that apply to heterosexuals," Kinman continued. "They are Americans. They've done nothing wrong. They are born that way. It's not a choice. It's not right that you have Americans who've done nothing wrong and you won't let them serve their country! It's wrong, just like the arguments against blacks were wrong in the fifties."

When I talked to him later, Volonino surprised me by being somewhat neutral on the subject. "I think women could serve on subs and I could command them," he said. The smaller fast-attack subs weren't roomy enough to accommodate them but the big Tridents were, he thought. But he also didn't think too many women were interested in serving aboard subs, particularly after they saw the cramped spaces. Juggling so many jobs with so few sailors aboard to perform them gave him headaches enough with all of them being males. Adding women to the mix would complicate it even more. "But there's no reasonable reason why women can't serve on subs," he said, with a caveat. "If it's done, it would have to be done right. It would take money and will. But right now there's no money and will to do it." On gays: "We've always had them serving on subs." As long as they kept their sexuality private and didn't chase other men on the boat he had no problem with their being aboard.

Levering finished serving dessert—for the noon meal it was a chocolate brownie with soft chocolate ice cream smothering the top—and the mood in the wardroom lightened some. The junior officers were getting used to Hunnicutt, who didn't seem as fearsome as they had imagined. As an icebreaker, Volonino told stories about the stunts Kinman and Hardee had pulled on patrols. Hunnicutt had already heard about Kinman.

They turned to sea stories. Hunnicutt was a naval history buff and enjoyed recounting famous sub battles.

"Pop quiz, Mr. Kinman," Volonino said at one point, turning to his young lieutenant. Volonino loved to throw out questions during meals. "What was the one thing the United States produced during World War II that was key to turning the tide of the war?"

Kinman squinted his eyes and thought and thought. "The one thing that turned the tide," he murmured to himself.

He finally gave up. "I don't know."

"It was merchant ships," Volonino answered. "The U.S. was able to produce them faster than Nazi

U-boats could sink them. World War II was a war of attrition, and the United States simply poured more war supplies into Europe and wore Germany down."

There was a lull in the conversation as the officers ate the last of their brownies and ice cream. Volonino finally broke the silence, turning to me.

"So what's the latest gossip on the White House?" he asked.

The latest gossip, the only gossip, topic A with this White House, was Monica Lewinsky. But I decided to take the high road.

"The air campaign over Kosovo is the biggest problem the White House is dealing with now," I offered, somewhat legitimately. It was a war, after all. "How Kosovo turns out could affect Clinton's place in history and Gore's chances for the White House. A lot of political futures are riding on this war."

"Why are we in Kosovo?" Kinman blurted out, as if it had been something that had troubled him for a long time. "I just don't understand it."

I recited the State Department's boilerplate. The United States had geopolitical interests in that part of the world. A stable Kosovo was key to stability in the Balkans. Turmoil in Kosovo could spill over into Albania, Macedonia, and Montenegro. Wider interests could be affected. Greece could come in on the side of their cultural allies, the Serbs, who were attacking Kosovars. Turkey could side with the province's ethnic Albanians, who were Muslims. Then you could have a real nightmare. Two North Atlantic Treaty Organization members, the Greeks and Turks, who had never gotten along, could start fighting on NATO's southern flank. "You could have the same type of instability in the region that sparked World War I," I ended.

"I don't think you can draw a parallel here with World War I," Hunnicutt said skeptically. "World War III wouldn't happen from this."

"Probably not," I agreed. "But just last month, Boris Yeltsin made rumbling noises that Russia might be drawn into the fighting over Kosovo if Washington didn't pay attention to his concerns." Madeleine Albright had taken the threat seriously enough that she phoned the Russian foreign minister and asked him if the alert status of his strategic rocket forces had been changed. They hadn't, the minister told her.

"What about you?" I asked the officers. Had they noticed Moscow's saber-rattling?

"It did perk up our ears," Volonino conceded. But the Tridents remained on the same alert they always were under, which was high.

"Diplomats at the State Department had a nickname for the day that Yeltsin made that threat," I said almost as an afterthought. "They called it World War III Day."

Silence in the wardroom.

Hunnicutt and Volonino looked at me as if I had told an offensive joke and shook their heads.

"How could they make light of such an awful possibility?" Hunnicutt muttered.

10 · Looking Outside

The Lord had blessed Reggie Rose. And there wasn't a day he woke up that he didn't thank the Almighty for what He had given him. Nor was there a day he didn't study his Bible intently and preach to others from his corner table in the crew's mess if they were willing to sit and listen. Yes, the Lord had been good to Reggie Rose, and he was grateful. He had a fine son, a wife he loved dearly, and on the USS *Nebraska* he was the best "on the sticks" of any sailor aboard. No one could drive this monster better than Reginald Dwayne Rose. No one. And he was proud of that. Proud that he had trained most of *Nebraska*'s planesmen, who sat before the two steering wheels in the control center—or "sticks" as the sailors called them—that moved the sub right or left, up or down.

And when the sub had a difficult maneuver, like the one it was about to make, he was proud that it was Petty Officer Second Class Reggie Rose whom the skipper wanted sitting at the seat for the inboard planesman. That was the seat in front of the right wheel at the control panels, which now controlled the *Nebraska*'s fairwater planes and the rudder.

Rose sat before the inboard stick now, his hands resting gently on the wheel, moving it ever so slightly

to make adjustments in course and depth. His eyes had
to keep darting about, to the digital and analog indicators on the panels in front of him, which showed the
depth, course, and angle of the sub. He had to make
sure that he or Jason Bush, the sailor sitting before the
outboard wheel to his left, didn't put too much pressure on their sticks one way or the other, which would
cause the sub to tilt and sail out of trim. It took a while,
but eventually a sailor developed a feel for the touch of
the stick and didn't overcompensate. The diving officer
sitting behind them wouldn't allow them to drift off
more than a degree or two in any direction. If they kept
the *Nebraska* sailing flat, straight, on course, and at the
specified depth, then everyone stayed happy.

It was 7:45 P.M. on Friday and the sun would soon be
setting on the water's surface 160 feet above them.
Volonino had already finished the afternoon intelligence briefing he convened every day in the officers'
study. Brady had reported that a Greek freighter was
sailing off the Florida coast. The only information
Naval Intelligence had on the merchant ship was its
registry. The other vessels anywhere near the *Nebraska*
were all U.S. warships, to which the navigation team
had already been alerted. They had drafted a one-page
"material status report" listing the equipment problems
discovered on the voyage so far. When the sun set, they
would raise the sub to periscope depth, poke the scope
and an antenna mast out of the water (the two would be
impossible to spot in the dark), and radio the status report back to headquarters.

Ganteaume had now taken over as officer of the
deck for the evening and would control the sub's ascent to periscope depth. He ordered Rose to turn the
boat around so the sonar dome at its nose could sniff

for sounds in the *Nebraska*'s baffles and make sure no sub or surface ship was following from behind when it went up.

Sonar found nothing behind them. The *Nebraska* was sailing at the edge of the continental shelf about seventy miles off the East Coast, where the depth dropped off to 1,800 feet or more. The Navy gave the Trident a huge trapezoid-shaped operating area there with no other U.S. submarines sailing inside it for the week. Ganteaume grabbed a phone overhead and called Volonino to ask for permission to go to periscope depth. Volonino granted it.

"All ahead one-third, steady course zero-two-zero," Gaunteaume ordered, clipping the phone back to its hook.

"All ahead one-third, aye, steady course zero-two-zero," Rose said crisply. You had to learn to repeat the commands quickly and accurately. Miscommunications could be disastrous. The officer of the deck told you to turn the boat one way and you turned it the other. Or sometimes an OOD ordered a turn and you knew it was the wrong turn to make, Rose had found. Those times, Rose had repeated the order, then said politely, "Officer of the deck, be advised that is the wrong way." He had saved several lieutenants' asses on the *Nebraska*.

Rose moved the wheel only slightly. He had a light touch. When he started on the sticks, he had been a nervous wreck, holding the wheel in a death grip, always worried about making a mistake. Figuring out how to move the planes was like learning to drive Dad's car. It took time to get used to the wheel. The basics could be picked up fairly quickly. A week of driving and a sailor usually felt comfortable at it. But it generally took an entire patrol before he had it mastered to the point that he didn't overcompensate manipulating the wheel or the diving officer didn't have to constantly prompt.

The *Nebraska* was more sluggish than Dad's car. Getting it to turn took time, particularly the slower it traveled. The planes also reacted differently depending on the depth of the sub and the water temperature outside. A sailor had to develop a sense of when to push or pull, when to let off. Rose and Bush, who manipulated the stern plane, tried not to fight each other with their steering. If one pushed his wheel in, the other might have to pull his back to keep the ship level. After ten years of driving the boat, Rose knew almost instinctively when to compensate for the plane movements his partner made.

An African-American, Rose looked younger than his thirty-two years. His dark brown skin stretched tightly over his thin frame. He kept his head nearly shaved, and his wide eyes practically glowed behind his gold-rimmed glasses. He spoke in the soothing tones of a preacher, which he surely could have been. His grandfather, cousin, and uncle were preachers. He had been born in Valdosta, Georgia, and reared there by his mother and grandmother. His parents had divorced when he was five. Valdosta was close enough to the Kings Bay base that he still drove home on weekends to visit family.

Rose had joined the Navy out of high school in 1985, wanting ultimately to be a doctor. But after training as a corpsman and spending a miserable two years pounding the ground with the Marines, he had decided to switch and become a submarine storekeeper, the other job he now held on the *Nebraska* besides driving.

He had no dread of accounting for supplies or keeping up with mounds of paperwork. He loved it, in fact. Rose was obsessively meticulous and neat. He kept a "tickler" by his side always, a weekly planner on which he jotted down every task he wanted to perform. Every hour of every day was mapped out carefully so every goal he set out for himself would be remembered and met. He backed up the paper copy of his planner on a computer disk as well.

Everything Rose did in his life or planned to do, big or small, he wrote down. That way you remembered it, he fervently believed. He was training his son, Reginald Rose II, to be like this. Though his son was only two years old, he read to him every night when he was home. One page per night from an English grammar book. Rose could already see results. Reginald the Second could already say his ABCs and recognize words on a page. He could even remember the names of all the relatives.

Rose was a born-again Christian, and he had done his best to be a hardworking steward. In port, he toiled part-time as an orderly in an emergency room, volunteered eleven hours a week in a convalescent home, off-loaded a truck once a month from five-thirty in the morning until noon for a feeding program. At Kingsland Church of Christ near the base, he served as an assistant minister. He loved standing in front of the congregation to tell what he had learned about Christ. "I am on fire," he liked to say.

Rose introduced himself to all the new sailors who came aboard the sub and invited them to drop by his table if they wanted to chat about God and the Bible. He didn't push his religion on anyone. If they ignored him, fine. And most did. This younger generation of sailors was not religious. Most never seriously considered why God had put them on this earth. Their goals didn't stretch far beyond getting qualified on the sub or putting in their four years for the educational benefits, he believed.

But some were receptive. A half-dozen sailors dropped by the table at one time or another for what had become Rose's own Bible study group. During the meal breaks, they would discuss how they interpreted the scriptures. Under way, he also kept the Bible's text on a computer disk.

Rose loved the Navy almost as much as he loved the church. He hoped to become an officer, then one day go to medical school to be a doctor. God willing, he wouldn't stop until he'd met that goal.

"Dive, make your depth eight-zero feet," Ganteaume continued with his orders.

"Make my depth eight-zero feet, aye," Shawn Brown repeated. It was his turn as diving officer, and he stood behind Rose and Bush.

Volonino had also slipped silently into the control center from the portside entrance. He wasn't announced as usual because the sub was ascending. Once Ganteaume had ordered the vessel to rise, the control center's men stuck to a strictly scripted set of reports until the Trident safely reached periscope depth.

Sailing at periscope depth was the least favorite thing a diving officer did. Brown was always nervous during the maneuver. Screw up and everyone on the boat knew it.

The danger was accidentally allowing the sub to broach the surface. It could happen easily if you weren't careful. Bush and Rose could start working at cross purposes. The sub could start drifting up. High waves on the surface could also cause the sub to broach, particularly when the waves came from the direction of the stern. The waves could suck you up in a heartbeat. The fairwater planes on the sail popped out of the water during the low part of the wave, then the wave slapped the planes at its peak. Everyone on the boat heard those damn waves slapping the planes.

And they let you know that you had fouled up. Submariners called it "getting your wings," and every diving officer had suffered the humiliation of broaching the ship. It had happened to Brown during a winter patrol when the waters were always rough. The sonar shack had called the direction of the seas wrong. Brown had ordered the planesmen to pull the sub up to a depth of eighty feet, and sure enough, a wave smacked him in the ass and sucked up the boat. For his gaffe, the crew presented Brown with a set of silver Air Force aviator wings.

If his planesmen were inexperienced, Brown would backseat-drive them the entire way to control the rate of ascent he wanted. He didn't have to with Rose or Bush. Brown just told them: "Take charge of your planes and make your depth eight-zero feet." That was the depth at which the periscope would poke out of the water when it was extended.

Rose knew how to get there. He pulled his wheel back to begin the ascent to eighty feet.

The lights were dimmed in the control room, which was now silent except for Brown calling out the depths he read from the digital and analog indicators as the sub rose.

"One-four-zero feet.

"One-three-zero feet.

"One-two-zero feet."

The sub hit a layer of cool water that made it more buoyant than Brown wanted, so he ordered James Penn, who sat to his left as chief of the watch, to begin flooding water into trim tanks. Penn, one of the sub's senior storekeepers, quickly began punching buttons and turning knobs on the ballast control panel.

"One-zero-zero feet," Brown continued. Rose kept the wheel pulled to his chest.

"Nine-five feet."

Rose pushed the wheel forward to begin leveling out the sub so it would settle at eighty feet.

"Nine-zero feet."

"Eight-nine feet." Rose kept the wheel pushed forward. It took time for the shifting planes to have their effect.

Ganteaume grabbed the two handles to the persicope in front of him and pressed his face to its eyepiece.

"Eight-five feet.

"Eight-four feet."

"Scope's breaking," Ganteaume announced quietly when he could see through his eyepiece that the periscope's lens had just cleared the water's surface.

"Eight-two feet," Brown continued. "Eight-one feet."

Rose pushed the wheel forward again, then back slightly, trying to set the boat at eighty feet.

Ganteaume circled with the periscope's eyepiece pasted to his face.

"No close contacts," he finally said. Nothing but darkening skies.

Quiet voices broke out in the control center. Everyone was always relieved when the man looking through the periscope reported there were no surface ships nearby that might collide with the sub. In the background, a speaker behind the number two scope began playing a nonsensical melody of beeps and chirps. It was produced by the electronic surveillance system sensor on the top of the periscope that now swept the skies for any radar emissions from another ship or airplane in the area. If the system detected a radar beaming from a ship too close, the speaker in control would let out a loud screech and Ganteaume would immediately send the sub into a dive.

"Seven-nine feet," Brown announced. Rose had nailed it pretty close to the depth they wanted. He shifted the wheel ever so slightly forward to inch the sub back down to eighty feet.

Ganteaume made another circle on the periscope. "Sea state one," he announced. The waters were calm outside. Brown wouldn't have to worry about a wave's yanking him up. Penn stopped pumping water on and off as the sub stayed level.

Almost imperceptibly, Rose shifted the wheel back and forth. The depth indicators shifted between 79.4 and 80.3 feet.

"Chief of the watch, raise number one multifunction mast," Ganteaume said.

Penn repeated the order and flipped a switch on his panel to raise one of the two thick masts behind the periscopes. They were used to transmit radio messages back to shore and receive position updates from GPS satellites orbiting above.

"Conn, radio," a speaker in the control center announced. The voice came from the radio shack next door. "In sync, VLF on number one multifunction mast." VLF stood for very low frequency, which was one way the sub could retrieve messages from land.

The ESM speaker kept up its silly melody. Rose kept giving his wheel tiny tugs and pushes.

"Eight-one feet," Brown announced. The *Nebraska* had dipped slightly. He whispered instructions to Rose, who pulled back just a little more. He was beginning to tire. The first time a sailor sat at the sticks, it was interesting. Weeks later it became boring. The driving had its difficult moments: when the sub traveled at periscope depth, for example, or when it sailed into and out of port and the planesmen had to react instantly to commands, or during high-speed operations when the sub fired torpedoes and turned and changed depths constantly. The wheel sometimes shimmied when the sub moved fast through the water. Otherwise, the driving could become tedious, particularly when the *Nebraska* was hiding on strategic alert, puttering along at three miles an hour to be more silent, and making no turns. Rose could sit at the stick for about an hour before he had to have a sailor relieve him so he could stretch and grab a cup of coffee.

"Conn, nav, GPS tracking complete," the radio speaker announced again. The navigation department had its fix from the satellite. "No longer require number one multifunction mast."

Ganteaume ordered it lowered.

Volonino walked through the forward entrance to the control center and stopped at Rose's seat to his right.

"Well, guys, you got it locked in?" he asked cheerily.

"Yes, sir," Rose answered nonchalantly. "Cruising around in an eighteen-thousand-ton ship is just a walk in the park."

"Bring on the Russian navy," Volonino joked back.

Rose smiled.

Twenty minutes later, Ganteaume had lowered the periscope and the *Nebraska* had sunk deeper into the ocean. The material status report had been radioed to Kings Bay without a hitch. Its paper copy would be on the squadron commander's desk the next morning. Maneuvering the *Nebraska* now became boring for a planesman. Straight lines, an occasional turn to stay in their operating box. Rose propped a foot up on a metal post, draped one arm over a knee, and steered with the other hand. He settled in for a long drive.

Other parts of the sub stayed busy. By 10 P.M., the radio shack located forward of the control center on the same level was in the middle of processing the night messages the *Nebraska* had downloaded from its communications buoy. Eric Liebrich stood over one of the two SID—standard information display—consoles at the far end of the radio shack. The green screen on this SID now showed a long list of messages the *Nebraska* needed to copy that had been sent out to the Atlantic Fleet's submarines.

This wasn't the postal system. Hundreds of messages came into the radio shack every day, many of them not for the *Nebraska*. The Navy's fleet headquarters found it easier to have one broadcast for all the submarines in the Atlantic. It meant that the *Nebraska* got to read everybody else's mail. The shack's radio receivers tuned in for the messages at regular intervals when the sub wasn't on strategic alert and poised to launch its missiles. When it was on alert, the messages were downloaded continuously.

Liebrich turned behind him to glance at another screen for the spectrum analyzer, looking for spikes in a white line dancing across it that indicated which of the shoreside transmitters or airborne command and control planes flying above were broadcasting. A

locked safe nearby was loaded with top-secret communications logs, code books, compact disks, and zip drives.
The roll of toilet paper used to wipe grease-pencil marks
off the plastic-covered communications status board
hung by a cord. Empty cups with brown coffee residue
sat in drink holders.

"The best way to describe my job is hours and hours
of boredom interrupted by minutes of hysteria,"
Liebrich said, almost to himself, as he stared at the
spectrum analyzer for signs of a broadcast. "We are the
keepers of all knowledge."

Liebrich was a supervisor in the radio shack. He had
joined the Navy fourteen years ago after high school,
not for the money or the educational benefits, but out
of pure unadulterated patriotism. Ronald Reagan had
inspired him. The Cold War was still being fought. The
Navy had a rich history and was still trying to build six
hundred ships to best the Soviets. Higher learning
could wait. Liebrich hadn't been interested in education just yet and probably would have wasted his father's money if he had gone to college after high
school. He saw submarines as his first calling.

Maybe he had a sense of duty because of the Ro-
mans. The one course he had enjoyed in school was
history. The Roman Empire became too big, its soldiers couldn't protect the borders, and barbarians dis-
membered it. Could the same happen to the American
empire? He didn't know. But at least he had his chance
to fight the Cold War, if only in its waning years. The
enemy had been real. So had been the risks with Soviet
and American subs secretly playing rabbit and wolf
under the seven seas. The best friends he had ever
made came during those tense times.

You couldn't find that kind of camaraderie now, he

lamented. The world had changed. The Soviet Union had collapsed. Thankfully, the U.S. submarine force hadn't broken down as the Russians' had. He still believed in deterrence. Americans didn't realize there were dangerous countries still out there like China and North Korea, and if they struck, the United States had to be prepared to strike back. But the urgency to perform, to be at the top of your game because the other side was, would never be the same, he realized.

Liebrich had talked his personnel detailer into assigning him to the *Nebraska* in 1997. It came as close to having that "old boat" quality of any Trident he had scoped out in the fleet. He had been on nightmare boomers where the skipper was interested only in his next rank and the sailors, chiefs, and officers were at one another's throats. The *Nebraska*'s crew wasn't as tight as the ones he'd served with when the Soviets still stalked the seas. But it came as close to being an old boat as any sub could in peacetime. It respected traditions, like making the nubs work hard to earn their dolphins and having sea daddies who mentored the youngsters. The officers and chiefs treated you like family and respected your technical expertise. If he ever went to war, God forbid, Volonino would be the type of captain he'd want to follow into it.

Liebrich had decided to finish out his twenty years in uniform. He was a petty officer first class now. He'd stay longer if he made chief. He thought he'd be a good one. He was proud that he had been a sea daddy for many young sailors. Tom Horner, who was sitting at the other SID console to his right, had been one of his charges. He had gotten him through the rough days when he first came aboard, had lent him a friendly ear, and, when it was needed, had sat him down for "one-way conversations" to keep him motivated. If Liebrich didn't make chief, that was fine with him as well. He'd retire at a young thirty-eight and perhaps teach. He was a semester short of earning a bachelor's degree. He wanted to be a historical archaeologist.

The right SID that Horner sat in front of monitored all the equipment in the room to make sure enough of it operated in sync to always receive messages. If the shack "lost sync"—like a car radio that gets out of tune when you're driving along and the music fades to static—the two operators might have to change antenna patterns or tell the officer of the deck in the control room to turn the sub to another course so they could pick up the signal again.

Horner was only nineteen, and he looked like a young Roddy McDowell. He was small and thin, with delicate features and an ever so slight smile that never seemed to leave his face. Horner had been an honor student back at Harbor Creek High in Erie, Pennsylvania, but his family had no money for college, so he joined the Navy after graduating. His enlistment was up in 2002 and he planned to get out, with enough money to finally go to college. He was already engaged, to Lacy, who was a year younger. Their backyards joined each other in Erie and they had gone steady since ninth grade.

Horner had always been interested in electronics—his stereo and CD player were always the loudest on his block—so he became a radioman. He had ranked at the top of the class in all his Navy schools. Horner believed that was why he was a petty officer third class now, at only nineteen. He stayed out of trouble, never drank, retained facts well, and kept his mouth shut.

He had learned how to play the Navy's game by his

third week of boot camp. When he came aboard the *Nebraska,* it hadn't taken him long to learn that on this "old boat" you were a second-class citizen until you earned your dolphins. The sailors without their dolphins were the first to be assigned the cleaning chores and the last to go on liberty. The A-gangers made you polish their shoes and fetch them drinks so they'd sign your qualification card. Sailors with dolphins would cut in front of you at the chow line. Hell, at sea, you couldn't even go into the enlisted lounge to watch a movie if you didn't have your dolphins. Horner worked furiously his first patrol, sleeping only three hours a night, so when the sub returned to port after a little over two months he had those dolphins pinned to his chest.

Between the two consoles that Horner and Liebrich manned sat a status alarm panel, a voice circuit box for radio, and special tape recorders for downloading weather information that was then fed into the Trident missile's computer just before launch. Behind the two men stood two rows of refrigerator-size consoles full of receivers, signal converters, demodulators, encryption and decryption devices, message processors, scanners, a spectrum analyzer, computers. The radio room had two computer systems: one electronically linking the antennas to the receiving equipment, the other processing the incoming electronic signals into readable messages.

The shack was set up so that a number of electronic ears operated at the same time. Liebrich and Horner had two of everything. Practically every piece of equipment had a backup. There were more than two dozen ways the radiomen could receive messages from shore. Two communications buoys, for example, could be strung out to float behind the submerged sub at about ten feet under the water, vacuuming very low frequency, or VLF, signals sent by STRATCOM and the Navy commands. Short wires attached to the buoys, called "pigtails," could also pick up high frequency, or HF, signals. The multifunction mast could be popped up at periscope depth for ultrahigh frequency (UHF) signals.

If the sub had to operate more covertly, still another wire antenna could be reeled out for two thousand feet to receive extremely low frequency (ELF) signals that penetrated deep into the water. The ELF signal came in agonizingly slowly, so the message consisted of only three-letter codes. The shack had an inch-thick book in its safe that could translate each trigraph, often sent as a bell-ringer to order the sub to sail nearer the surface so it could pick up a lengthier message on another frequency.

The Navy never wanted the *Nebraska* or any of its other Tridents to be out of touch if the order had to be given to launch the missiles. Most of the equipment crammed into the shack was there to receive messages rather than to transmit them. Listening was more important than talking. In fact, Liebrich, Horner, and the other radiomen ended up being much better at receiving messages than they were at sending them, simply because they got so little practice transmitting. On patrol, a ballistic missile submarine beamed out few messages for fear the enemy could use signal direction finders to pinpoint its position. Only Volonino could approve a transmission. Tags were even placed over buttons and switches used for sending out signals, warning sailors not to power them accidentally.

Listening for the "EAM" was the radio shack's one

mission in life. "Everything else is cake," Liebrich liked to say. The EAM was the emergency action message, the order that came from Strategic Command to launch. "Nuclear Armageddon begins right here," he also liked to say. A console in the shack chirped a high-sounding beep and an amber light flashed on its screen when an EAM arrived. In the control center behind the shack, another red light flashed near the periscope and an alarm blared that sounded like a warble.

The EAM always came encrypted in a top-secret code that scrambled the message into unintelligible four-letter groups. The radiomen decoded its heading, which identified the sender, what sub was supposed to receive it, and what kind of message it was. If it was an EAM for the *Nebraska,* teams of officers were quickly summoned to the shack to decrypt the rest of the message and verify that it was an authentic order to launch.

The shack received hundreds of practice EAMs so the crew could rehearse the procedures for quickly launching the missiles. Liebrich believed that if he ever had to process a real order, it would mean deterrence had failed. No American President would launch these missiles out of the blue. It sounded like brainwashing, Liebrich knew, but he had confidence in his country and its leaders that they knew these awful weapons couldn't be used capriciously. The world would have to have gone to hell and the United States itself would have to be under attack before these weapons of mass destruction would be launched. He'd have no qualms about firing back then.

Even the practice EAMs always seemed to unnerve Horner just a bit. He didn't know how he would feel if a real order came in, but he was sure he never wanted to find out. A real message would mean the *Nebraska* would be killing a lot of people—an awful lot. Horner thought he was a strong-willed person. He'd follow the order. But he was certain he would be "traumatized" for the rest of his life—if there was life after a launch.

Horner and Liebrich began ripping paper off the printer next to their SID consoles and stamping routing indicators on sheets. The Navy's fast-attack subs had e-mail, so incoming messages could be delivered electronically to computers throughout the sub. The Tridents still moved paper. The radiomen roamed the *Nebraska* with clipboards full of messages, often jostling awake sleepy officers in their racks at all hours to deliver them.

Some messages came in doubly encrypted with intelligence or orders so sensitive only Volonino could see them. He would be called to the radio shack, where he'd sit in front of a special decryption box and printer to decode the report personally.

But at the moment, the printer was spitting out news stories and the baseball scores. The Navy didn't want its Trident captains, or any of their crewmen for that matter, kept in the dark about what was happening in the world above them. So every week, headquarters broadcast about four single-spaced pages' worth of news, sports, and business reports to the *Nebraska* over an administrative channel. Just the day before, a radioman had rushed into the control center with the bulletin that NATO warplanes attacking Serb sites to stop the slaughter in Kosovo had accidentally bombed the Chinese embassy in Belgrade. The report had sent a chill through control. China had nuclear weapons.

Most of the crew lived for the paragraphs in the news summary that gave the ball scores. Liebrich and Horner were more interested in the stock reports.

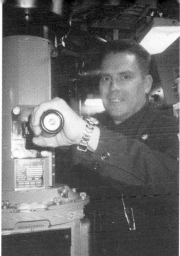

Commandeer Dave Volonino stands before one of the *Nebraska*'s two periscopes. Every time he sets out on patrol, he commands the sixth-largest nuclear power in the world.
(Douglas C. Waller)

Lieutenant Brent Kinman gives the treadmill a workout on every patrol. To combat bulging stomachs from the heavy meals and months underwater, the Navy has installed exercise equipment in all its Tridents.
(Douglas C. Waller)

Chief of the Boat Dave Weller *(center)* relaxes in the crew's mess during halfway night. Weller keeps close tabs on crew morale, and halfway night is one way the crew lets off steam. *(Ben Dykes)*

Lieutenant Commander Harry Ganteaume takes a rare break in the officers' study. As the ship's engineer, he has one of the most demanding jobs aboard a modern nuclear submarine. *(Ben Dykes)*

Lieutenant Junior Grade Ryan Hardee is
a prankster. But among the junior officers,
he's considered a skilled tactician during
battle drills. *(Ben Dykes)*

Lieutenant Commander Alan Boyd *(right)* and Lieutenant Al Brady stand
on the bridge, keeping watch for other vessels while the *Nebraska* is
surfaced to repair a sonar cable. *(Douglas C. Waller)*

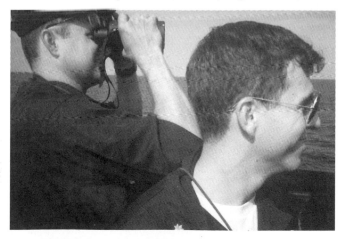

Lieutenant Junior Grade Chad Thorson is reunited with his wife, Kyung, the day the *Nebraska* comes home. The long patrols can be just as hard on wives as on the crewmen. *(Douglas C. Waller)*

Chief Robert "Doc" Philbin perches on the examining bed in the sub's pharmacy. During the patrol, Doc faced a rash of illnesses, including a case of appendicitis. *(Douglas C. Waller)*

Lieutenant Al Brady *(second from left)* supervises sailors at the control center's plot tables during a maneuver. The ship's navigator, Brady plots the top-secret course the Trident will sail during its patrol. *(Ben Dykes)*

Lieutenant Steve Habermas, the *Nebraska*'s communications officer, conducts Sunday lay services. Using lay leaders, the sub holds Catholic and Protestant services each Sunday. Many of the submariners are deeply religious. *(Douglas C. Waller)*

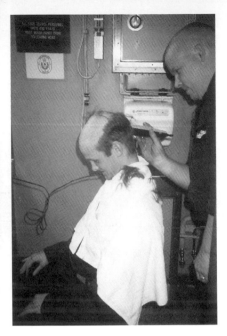

Ensign Ray Chesney sits for his head shaving at the beginning of the patrol. As the *Nebraska*'s "bull ensign," Chesney must endure endless pranks during his first assignment in the submarine service. *(Douglas C. Waller)*

Petty Officer Second Class Chris Wilhoite poses with a pirate's eye patch. He came down with appendicitis during the patrol. Doc pumped him full of antibiotics and saved his life. *(Ben Dykes)*

Three missile technicians in the missile compartment: *left to right*, Petty Officer First Class Mark Lyman, Petty Officer Third Class Greg Murphy, and Petty Officer Second Class Keith Williams. The *Nebraska*'s twenty-four strategic missiles carry twice as much explosive energy as was released by all the conventional weapons in World War II.
(Douglas C. Waller)

Chief Marvin Abercrombie stacks the chips for casino night during the patrol. Suffering from a kidney ailment, Abercrombie was Doc's other major medical problem during the patrol. *(Ben Dykes)*

Ensign Mark Nowalk *(seated)* learns the duties of the diving officer under the tutelage of Chief Shawn Brown, who also directs the two planesmen: Petty Officer Third Class Jason Bush *(left)* and Petty Officer Third Class Quentin Albea. *(Ben Dykes)*

Commander Volonino *(left)* and Lieutenant Kinman lean back as the submarine dives during "angles and dangles." The crew sends the submarine porpoising up and down to limber it up for battle. *(Ben Dykes)*

With oxygen masks and protective hoods covering their heads, crewmen in the control center practice a war game at the same time that they are fighting a simulated fire aboard the sub. *(Ben Dykes)*

Sonarmen crowd into the sonar shack, watching the green waterfalls on the "stacks" and listening for the sounds of the enemy, during practice maneuvers against another sub. *(Ben Dykes)*

Petty Officer Second Class David Smith *(left)* and Petty Officer Third Class Alfredo Donis check out one of the sub's torpedoes. If it has to, the Trident can become an attack sub, firing torpedoes at other vessels.
(Douglas C. Waller)

During a simulated fire in the missile compartment, the "nifti" operator, carrying a thermal imaging device, looks for hot spots, with a hose team squatting behind him. Blue hospital hair nets cover the oxygen masks of the hose men to simulate vision obstructed by smoke.
(Douglas C. Waller)

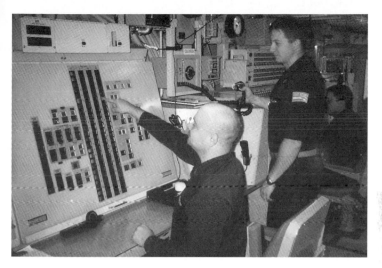

During a simulated missile launch, Lieutenant Fred Freeland prepares to pull the trigger while Petty Officer First Class Mark Lyman punches buttons on the fire control console. *(Douglas C. Waller)*

Using the scoring for a Scottish dart game, Chief Stacey Hines keeps track of the missiles being fired during a simulated launch. The Trident crew drills constantly so that it will be prepared to launch at a moment's notice. *(Douglas C. Waller)*

Nicknamed "the house," this huge compartment contains long rows of giant tubes holding the *Nebraska*'s D-5 ballistic missiles. The tubes keep the missiles in a pristine state, ready to be fired quickly. *(Ben Dykes)*

A sailor inspects the seven-ton hatch that covers the tube for missile number ten. A thin fiberglass dome that also covers the missile explodes away when the missile is fired out of the tube. *(Douglas C. Waller)*

The security team draws weapons for a drill to test its response to a terrorist threat aboard the sub. The men are under strict orders to protect the missiles at all costs, even if that means hostages are killed. *(Ben Dykes)*

As the *Nebraska* nears land, a hawk perches on one of the masts and is spotted through the periscope. *(Ben Dykes)*

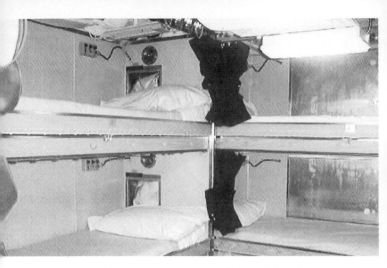

Living quarters are cramped aboard the sub.
Beds are stacked three high for the crew.   *(Ben Dykes)*

Petty Officer Second Class Jason Duff, one of the cooks, prepares lunch
in the galley. Meals are respites from the boredom of the long patrol. A
good cook can be important for morale.   *(Douglas C. Waller)*

The crew gathers in the enlisted mess for pizza and ugly-tie night. Commander Volonino, in the chef's hat, is one of the cooks. *(Ben Dykes)*

Petty Officer First Class Eric Liebrich *(second from left, in sunglasses)* normally works in the communications shack processing launch orders. But on halfway night he's the emcee. The men celebrate making it halfway through the long, lonely patrol with a party. *(Ben Dykes)*

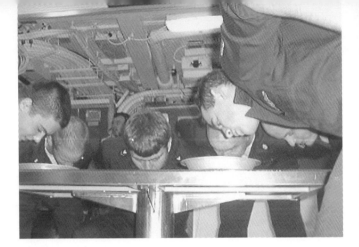

Seaman Ryan Beasley *(far left)* joins other sailors in a Jell-O–eating contest during halfway night. Raffles and games are held all day during the halfway festivities. *(Ben Dykes)*

During a rare "steel beach" when the *Nebraska* surfaces, Commander Volonino and Chief of the Boat Weller serve hamburgers to the crew. *(Douglas C. Waller)*

Liebrich, who was the sub's financial adviser for seamen with money problems, had a fat portfolio of
stocks and mutual funds. Horner had been playing the
market since he was fourteen. "Never bet a radioman
on sports scores," Liebrich said, however. "We always
see them first."

The access became both a perk and curse. Liebrich,
Horner, and the other radiomen were the first on board
to see practically every message that came to the sub.
They had the highest security clearances of any crew
member, some of which were so sensitive that even the
fact they possessed the clearance was kept a secret. It
meant a lot of face time with the captain delivering
top-secret reports. But processing radio messages also
gave them the first glimpse of news that the crew desperately craved, such as when the sub might dock at
ports, the birth of a sailor's son or daughter, or an illness or death in a family.

The rule was always that Volonino delivered the
good and bad news personally. Or decided what got
delivered. Sometimes a family tragedy was kept from a
crewman working in, say, a sensitive job with the nu
clear weapons for fear it might make him emotionally
unstable. Horner could be having lunch with a friend
whose wife he knew had just miscarried their baby and
he couldn't say a word to him about it because the sub
was on strategic alert. That could eat you up inside, he
had found. But radiomen were under strict orders to
keep their mouths shut. No gossip.

Liebrich and Horner began tearing apart the computer paper and stamping pages with routing indicators. Suddenly the EAM alarm began beeping from the
console. An emergency action message was coming in.

"Now comes the hysteria," Liebrich said, but not

looking too hysterical. He pulled a decryption manual from the safe and began flipping pages. With the manual folded open in his arms, he looked to the console screen, at the four-letter combinations in the beginning of the message, then matched them to the codes on the pages he turned.

"It's a STRATCOM test," Liebrich finally said, dropping the decryption manual back into the safe. "It's of no concern." STRATCOM was always sending out test messages to different subs. The *Nebraska* didn't have to pay attention to this one.

Liebrich and Horner continued stamping pages. It was nearly eleven o'clock at night and they wanted to get the messages delivered before midnight rations. "Cook's surprise tonight," Liebrich said. A secret he didn't care to know ahead of time.

11 • The Hostage

Volonino walked into the wardroom at 5:40 A.M. on Saturday looking jaunty with his blue cotton turtleneck under his freshly pressed blue poopie suit. A fruit plate had been set at the head of the table for him. The cooks never knew exactly what he would eat each morning for breakfast, but after a while they realized that he liked to start it with a fruit plate, so they always had one ready. Though hours seemed to lose their relevance on patrol—someone was always up working, and day and night were created artificially with the lights—breakfast was served at an ungodly early hour, 5 to 6 A.M. sharp. Arrive at 6:01, after the galley had stopped serving, and you had to fend for yourself with only a sweet bun left on a shelf and coffee from a pot that was always full and steaming.

Frank Levering, who had heard Volonino come through the wardroom door, quickly came out from the aft pantry door and stood silently by the front of the table as the captain settled into his chair and decided what he wanted with his orange and grapefruit slices. This morning it was a bowl of cereal and three strips of bacon.

Volonino's eyes were bloodshot from lack of sleep the past week. All the officers were bleary-eyed from the heavy workload. But Volonino, more than anyone, reveled in it. That was the book on him, according to the other officers. He loved the grind of sea duty, loved driving the boat around the ocean, loved being a ship's captain. Even Volonino admitted it. He knew, of course, that the Cold War was over. The world was much more to his liking than it had been forty years ago when communists intended to change it. He also knew it didn't make sense to continue spending the large sums America had been spending before on defense. But it also didn't make sense to do away with the Tridents. Practically every town in Europe had a monument to the folly of conventional deterrence, a statue or graveyard honoring the fallen from World War I and II.

The way Volonino saw it, nuclear deterrence was as important now as it had ever been. In fact, more eggs had been placed in the strategic nuclear basket. Europe had long been stripped of the huge conventional army the United States had once kept there. The U.S. Navy was the only force left that still showed its muscle in the region. American tanks would no longer be the first line of defense against an attack. Nuclear weapons would.

On his first patrol with the *Nebraska* in 1997, the admirals had dispatched the sub to the Mediterranean Sea and ordered it to pop up at a Greek port and entertain a bunch of NATO bigwigs, all in full view of Russian spies. During the Cold War, a Trident would never have dared to sail into a confined ocean like the Mediterranean, crowded with ships it might bump into, much less surface for all the world to see. But the admirals wanted to show NATO and Moscow that even now that the Cold War was over, the United States still had the might to defend its allies.

At six o'clock on the dot, Chad Thorson entered the
wardroom and stood erect at the front of the table with
a steno pad opened in one hand as if he were about to
sing in a choir. He had just finished his six hours of
duty in the engine room as the engineering officer of
the watch, but he had one more ritual to perform before
he could fall into his rack for a few hours of sleep.

There were two officers who ran the boat from hour
to hour when Volonino was preoccupied with other
chores and not on the conn: the officer of the deck,
who was responsible for the entire ship and directed its
movements from the control center, and the engineering officer of the watch, who stayed in the back with
the propulsion plant. When they became qualified—
which took a while, because these jobs were far more
difficult than just being night watchmen—officers divided up the duty so each served a six-hour shift. And
after every shift, the two officers tracked down
Volonino and reported what had happened during their
duty time.

It wasn't a casual update. Every commanding officer
had his own preferences for how a watch officer's relief report should be delivered. Volonino had a very exact format for the monologue, with a specific list of
items he wanted reported each time—succinctly, no
rambling, so the watch officer didn't waste his time.
Many conversations subordinates have with a submarine captain are structured that way, for an important
reason. In combat or emergencies when the boat suffers casualties, a skipper finds it easier to mentally
process and act on a report if he hears it the same way
every time.

New officers always had trouble reciting their lines
properly. It took several months of uncomfortable mo-

ments with Volonino, stuttering out their reports, before
they learned to be crisp and prepared in their presenta-
tions. There were do's and don'ts, the rookies painfully
found out. Always gird yourself for a grilling. Never—
repeat, never—report a problem on the ship to the skipper unless you have a plan in the next breath for solving
it. Even if it's a dumb solution, at least offer something.

Much of Volonino's day was spent reading or listening to departmental reports. A watch officer's relief report was one way—and an important one—for him to
stay updated. The officer of the deck, for example, always recited all the contacts the sonar shack was tracking so the commander would know if any vessels were
sailing too close to the sub.

But Volonino also used the relief reports as a teaching tool. For example, he always knew the sub's position, along with the depth, course, and speed at which
it sailed. Boxes hung all over the *Nebraska* with the
digital readouts, and Volonino kept one eye on them
everywhere he went. He didn't need the officer of the
deck to tell him where the sub was. But he required
him to include the position update in his report because
he wanted the young lieutenant to know that this captain considered it important to be current at all times. If
the lieutenant came unprepared and tried to bullshit
with a made-up position report, he'd jump on him like
a lion.

Volonino also threw a hardball question at every
watch officer who stood before him. This was always a
graduate-level query on any aspect of the ship, perhaps
something the lieutenant would see on a future qualification exam. If the officer was sharp and had some operating years under his belt, he stood a good chance of
knowing the answer off the top of his head. The

younger ones usually had to look it up and report back to Volonino within an hour.

Thorson nervously began his recital, rattling off details on the operating mode of the reactor, its current power level, the status of the electric plant, the results of the latest chemistry samples taken from the steam plant and reactor plant. Volonino tilted his head to the left and nodded after each item.

Then he fired his first hardball, a question on the reactor's operating procedure. Thorson knew the answer—or at least thought he did—and fired back. Volonino thought it was wrong.

"That's a lookup," he said tersely.

"Yes, sir."

"Go ahead."

Pulling out another sheet of paper from his pocket, Thorson turned to the status reports for the repair work under way in the reactor section. There had been a minor problem with the reactor start-up in port, and technicians were still investigating. Volonino interrupted Thorson's monologue to grab the phone and call the reactor section for more information.

Volonino fired another hardball.

Thorson had no earthly idea what the answer was. "That's another lookup," the young man said bashfully.

"Okay, what else?" Volonino said curtly.

"I don't have anything else unless you have further questions," Thorson said, hoping Volonino had run out of hardballs for now.

"I have nothing else," the skipper responded.

Thorson laid a report form down on the breakfast table for Volonino to initial. Then he hustled out of the wardroom to find the answers to the two questions.

Volonino enjoyed this reporting ritual. The pop

quizzes were his way of taking young officers under his wing. Mentoring was a tradition in submarines, and Volonino was sensitive to tradition. Sailors and officers sang the University of Nebraska fight song during the ceremony when they were awarded their dolphins. He kept a ship's Bible, which all the crewmen who reenlisted signed and then placed their hand on to swear allegiance to their country. Each man then received a *Nebraska* pen, an American flag, and a copy of the Constitution. Tradition was a glue that bound the men together. "It defines who we are," Volonino said.

Thorson didn't need an hour for his research. He returned fifteen minutes later, as Volonino finished the last strip of bacon, with a fat reactor manual cradled in his arms. The answer to the second question, the one he hadn't known, he spit out from memory. For the first question, he dropped the manual on the breakfast table, opened it to the middle where he had a yellow slip stuck to a page, and pointed to a paragraph.

Thorson had given the correct answer to the first question. Volonino read the passage in the manual, which proved that the lieutenant junior grade had been right and the commander wrong.

"Okay, very good," Volonino said, looking up with a smile.

Saying nothing more, Thorson closed the manual, scooped it into his arms, and marched out of the wardroom beaming.

"Now he'll go around and brag to everybody about how he just thumped the CO," Volonino said, chuckling.

Bob Lewis strolled aimlessly along the starboard passageway of the missile compartment on the second

level, his hands in his pockets, gazing this way and
that, for no particular reason. That was odd for Lewis.
Ever since the thirty-five-year-old chief took over the
A-gangers, he had been running around the sub like a
man with too many jobs and too few hours to complete
them. Now it was 9:15 A.M., one of the busiest times
for the *Nebraska,* and Lewis acted as if he were out for
a Sunday walk.

Bob Lewis had seen a lot in the Navy. He had sailed
on three submarines and on a tender ship that repaired
subs. He had instructed every kind of sailor at the Trident Training Facility at Kings Bay. He had spent three
years as a guard at the Navy's brig in Norfolk, Virginia,
the long-term prison where the service kept its most
dangerous dregs—murderers, rapists, spies, habitual
thieves, the sailors who would spend the rest of their
lives behind bars. He supervised a machine shop where
they worked. They were industrious enough—there
wasn't much else for a lifer to do in the brig—but
Lewis had a healthy respect for the evil in them. He
knew all their tricks, all the moves they could make to
overpower a guard in an instant if he wasn't careful.

Lewis ambled past the long row of missile tubes on
his right and then Murphy on his left. Murphy was preoccupied with readings on the missile compartment's
control and monitoring panel, called the CAMP, which
tracked the temperature and humidity in the tubes. The
guard had walked to the forward end of the compartment and was out of sight.

Twenty-four hours a day, the missile division had
two sailors roaming the four levels that housed the
tubes. Each had a nightstick strapped to his belt and a
whistle, which he would immediately blow if he spotted anyone acting suspiciously around the rockets.

Whenever technicians had to work inside a tube to make a repair, the missile compartment was sealed off and armed guards were posted.

The crew was always alert to outsiders who might try to sneak aboard. When the *Nebraska* was tied to a pier, two armed guards stood at the end of the gangplank, where it rested on the top deck in front of the hatch opening used to climb down into the sub. Few private cars were allowed to park on the dock near the sub. Civilians who were allowed inside the vessel for visits were always watched closely.

There was also the threat of an enemy within.

Lewis walked to the back end of the missile compartment near a machinery section, where he saw Jason Lawson. A native of Atlanta, Georgia, Lawson had only a slight southern accent. He had been on the sub for less than a year working as a junior technician in the missile division.

"Hey, Lawson, there seems to be some oil leaking over there," Lewis said, pointing to one of the panels near the stairs to other levels. "How about checking that out."

Lawson bent over and looked down at the deck, perplexed because he didn't see anything.

In an instant, Lewis stuck his hand into his right pocket and grabbed a knife. He reached down and wrapped his left arm around Lawson's neck and yanked him up. He had the knife pointed at Lawson's throat.

"I can't take it anymore!" Lewis began shouting at the top of his lungs as Lawson stood frozen in place. "I can't take it anymore!"

A sailor who was about to walk up the stairway from the lower level saw Lewis holding Lawson at knifepoint above and scampered back down the stairs.

"I've got a knife!" the chief shouted at the fleeing sailor. "You'd better get somebody here to talk to me!"

The knife was made of cardboard. Lewis was pretending to go berserk and Lawson was a willing hostage. Once a month while it was at sea, and every week when it was in port, the *Nebraska* crew held a security drill. The Navy was paranoid about any harm coming to the Trident's missiles, so Volonino and his drill coordinators spent hours dreaming up ways a terrorist or a deranged crewman might attack. In one drill, fake sticks of dynamite were wedged between the missile tubes while the sub was submerged, and the crew had to find a way to defuse the explosives or surface the ship to throw them overboard. The week before when the sub was still tied to the Kings Bay pier, the crew had pretended it had docked at an overseas port and had to cope with a foreign visitor on board for a tour suddenly holding the ship hostage.

One of the roving guards with a nightstick approached from the right side of tube twenty-four, which was near the back end of the missile compartment where Lewis stood with Lawson.

"I've got a knife," Lewis shouted at him. "I'm not going to sea. Get away from me!"

The guard backed off.

"Security violation in the missile compartment," the sub's speaker system began announcing. An alarm began bonging. Missile compartment guards quickly posted themselves at entrances to the compartment to prevent anyone else from stumbling into the scene.

At the forward end of the submarine, on the same second level, sailors assigned to the sub's security team lined up at a weapons locker near Volonino's stateroom and began donning flak jackets and strapping holsters and gas masks around their waists. Sean McCue, one of the missile division chiefs, stood at the locker passing out shotguns and pistols. He whispered to each sailor as he handed him a weapon: "Deadly force is not authorized. This is a drill." Otherwise the sailors would assume the crisis was real and shoot to kill.

Several doors down at the officers' study, Brady and Tremayne rushed in and ripped the plastic cover off the conference table in the center of the small room. Underneath the plastic, covered with Plexiglas, was a line diagram of the entire sub with charts and emergency checklists arranged on the side. The two long walls of the small study were lined with locked cabinets and combination safes. Built into the entrance bulkhead and the one at the other end were Naugahyde couches. The two officers began strapping phone sets around their necks and testing communications circuits while marking on the Plexiglas with grease pencils at the same time. The study now became the command center for dealing with the hostage crisis. Brady and Tremayne used the line diagram on the table to pinpoint the trouble spot and direct the security team members to it.

"Perimeter control has been established," Brady radioed to the control center. "The danger area has been cordoned off."

One floor up, at the conn, Volonino sat at his captain's chair with his legs crossed and propped up on the lip of the periscope to his right. He looked at his watch, checking the minutes it took for the security team to reach Lewis. They were taking too much time. During these kinds of emergencies, it was even more important that Boyd be with the response team and Volonino remain at the conn. It reduced the chance of both the commanding officer and his executive officer being killed by the terrorist.

During the last patrol, Volonino had tested the crew on how it would react if the most experienced person aboard tried to prevent the missiles from being launched. Announcing on the sub's speaker system that he had been gripped with a sudden attack of pacifism, Volonino stole one of the special keys needed to fire the nuclear weapons and scrunched between two tubes in the missile compartment where no one could see him, with a phone that he had also filched.

Then, in a disguised voice, he began sending out false sightings of the skipper over the speaker system. For almost an hour he had a delightful time listening to the footsteps of security teams racing up and down passageways every time he gave a false report, like little children chasing a soccer ball. The irritated sailors eventually wised up. One finally found him between the tubes and dragged him out.

The security team now had Lewis surrounded. Gregory Turner, a machinist, stood at the stairs leading up to the second level of the missile compartment and could see Lewis, who had backed into a corner with his hostage. Calvin Ireland, one of the senior missile technicians, crouched down on one knee beside tube twenty-four on the right with a perfect bead on the chief. Paul Dichiara and Jose Victoria, two other missile technicians, had crept along the opposite passageway with .45s strapped to their waists. Don Katherman, still another senior missile tech, stood in the space between the two rows of tubes with a dead aim on Lewis. The two sides were in a standoff.

Katherman began the negotiations to keep Lewis preoccupied as the other men maneuvered into better positions. "What's the problem?" he shouted.

"One more person walks through that passageway, I'm cutting his throat," threatened Lewis, who could see the team trying to surround him. "You've pushed me over the edge. I want the boat turned around!"

"We're working on that," Katherman said.

"I want to go back in port," Lewis continued, his arm wrapped across Lawson's chest.

"We're talking to control right now to have them turn the boat around," Katherman lied. "What are you going to do when we get to port?"

"I'm getting off," Lewis answered. "I want to go home."

"We're working on that," Katherman repeated.

"I don't think you can do it," Lewis said skeptically. "I want somebody here important enough that they can tell me they're really turning this boat around. If it comes from the CO or XO, I know it's the truth."

Katherman knew neither Volonino nor Boyd would negotiate directly with Lewis, so he kept stalling for time.

"Why do you want to go home?" he asked.

"I got stuff I got to do when I get back," Lewis answered. "I'll go to sea with you the next time."

"Anybody at home who can do it for you?"

"Nope, I got to do it myself."

"Let him go!" Dichiara shouted from the starboard passageway and aimed his pistol at Lewis so he could see it.

"I'll cut his throat," Lewis shouted back and swung Lawson around so he'd block the bullet. "What are you going to do, shoot him?"

"He's disposable," Dichiara said without emotion.

Unfortunately for a hostage, it was the truth. The security men didn't want an innocent person to die, but they were also under ironclad orders to protect the missiles at all cost. They would kill a hostage if the terrorist tried to use him as a shield to gain access to the nuclear weapons.

"Come on, chief, let's go," Katherman said, trying to sound more conciliatory. "We're maneuvering the boat as we speak."

"I don't think so," Lewis said warily.

"I'm going to shoot you if you don't give up," Katherman finally said with certainty to his voice.

Lewis thought about it for a few minutes, then dropped the hand that held the cardboard knife to his side. He let go of Lawson.

"Get on the deck, chief!" Katherman barked, his pistol still pointed at Lewis. "Put the knife down and lay down on your belly."

Lewis did as he was ordered and pushed the knife away.

"Okay, chief, get back up on your knees and put your hands behind your back," Katherman said. He simulated keeping his .45 pointed at Lewis while Victoria handcuffed his wrists with a plastic restraint.

"I'm going to take him to the wardroom," Victoria said, holding Lewis by one of his arms and keeping a pistol pointed at his back.

Katherman nodded.

Other chiefs would interrogate Lewis in the wardroom to make sure he had acted alone and there were no other conspirators they had to deal with on board. The sub had no brig. If Lewis had been a real hostage taker, he would be put in his bunk, where a guard would watch him twenty-four hours a day until he could be taken off the boat.

12 • Sounds of Silence

Ryan Beasley pointed his flashlight at the ceiling of the command and control center Saturday afternoon. With his other hand he grabbed a metal post, because the room was tilting forward and he didn't want to tumble out its front door. Beasley didn't know exactly why he was looking up. The petty officer who had given him the flashlight had just told him to shout if he saw any drops of water falling when he shined the light at the ceiling. The officers had nosed the sub down into a deep dive, and Beasley hung on for dear life as he squinted at the spaghetti of pipes, tubes, and wires covering the overhead.

The Navy would only publicly acknowledge that its Tridents sailed as deep as eight hundred feet. The sub actually could travel much deeper to escape the enemy, although how deep was kept secret. But in the first week of its patrol, the crew tested how the *Nebraska* reacted in a deep dive. After sitting in port for months, the sub could be out of shape when it was time to withstand the terrific pressure on its hull from a deep submergence. So now the control center's officers pointed the vessel down for a test, stopping every hundred feet to see if Beasley and other sailors posted around the boat spotted anything leaking from the walls or ceilings.

Beasley also wasn't quite sure how he had gotten on this sub, much less what he was supposed to do when he was here. Just nine months ago, he had been running a gas station in Concord, New Hampshire. He had decided to quit pumping gas and join the Navy so he could learn how to sail a boat. He had always dreamed of owning his own sailboat and sailing around the world. But among all the forms the Navy made him sign when he enlisted, there must have been one that put him in submarines. Beasley didn't know which form it was, but here he was, diving deep into the Atlantic Ocean in a Trident. And he didn't think he'd learn much about sailing on top of the ocean as long as he stayed under it.

Oh well, Beasley decided. He'd try to make the best of it—although that was easier said than done on this sub. Ryan Beasley was twenty-one years old, but he looked like a precocious teenager barely old enough to drive. His tanned face was smooth, with hardly a whisker, his short black hair sprouted in different directions, and he had a bright-eyed naive look that seemed to invite harassment. It had been that way for him in high school. He loved art classes and had never been interested in sports, except for a little fencing. He was a quiet sort who wanted to keep to himself. But everybody knew him in high school. Beasley had a bad habit of standing out—usually because he blurted something at the wrong place or the wrong time.

And now in this sealed tub, he seemed again to be standing out in ways he didn't want to. Beasley had come to the *Nebraska* as a "striker," the designation for young sailors who were sent to the sub immediately after boot camp so they could work in the different departments and then decide the specialty they wanted to learn in the service's advanced schools. Beasley had latched onto Doc, to begin apprenticing as a corpsman. The Navy wasn't going to teach him to sail and it certainly didn't have any jobs here drawing pictures with charcoals. So if he became a corpsman, maybe he could find a job as an ambulance driver after the Navy—until he decided what he really wanted to do with his life.

The Navy had another reason for sending strikers to the boats, Beasley soon realized. They were the labor pool for the menial jobs. So far, all he had done was clean things. And he still had a lot to learn about the rules of engagement here. Beasley wasn't mechanically inclined, and whenever he operated equipment it seemed like some petty officer was always barking at him. The older A-gangers, who usually supervised the strikers, liked to tease the young kids. When Beasley teased back, he quickly discovered that that wasn't done. The old-timers dished it out. But a striker who joked back was branded a smart-ass and given more floors to scrub.

As the sub settled to its lowest depth, Beasley luckily didn't see any drops on the ceiling, so he kept quiet. The *Nebraska*'s hull actually compressed about a foot under the enormous pressure from the water outside. That was why the four decks inside the sub were hung from the hull at a distance, so when the outside of the vessel shrank in deep water the floors didn't crack. The deep submergence proved uneventful, so the officers ordered the planesmen to pull the sub back up to 160 feet. Beasley hung on to the metal post again as the room tilted up.

Kinman finished his last mile on the treadmill, pounding it with his feet and panting. Sweat stains covered his white T-shirt, which had an image of the *Nebraska* printed on it. "Club Red," as the sailors called the exercise area in the back of the missile compartment, was Kinman's refuge every day. The crew could run on the long passageway surrounding the missiles on the top level, but this was difficult because the path was narrow and you had to dodge or duck metal objects along the way. Most of the men visited Club Red, which had two stationary bikes, two treadmills, a weight set with mat, and a Nautilus machine. Doc pestered all the seamen to exercise so they wouldn't become fat blobs from the heavy meals and sedentary living. Kinman gave the equipment such a workout that Volonino had ordered him to run on the right treadmill so it would be the only one that broke by the end of the patrol.

Winded, Kinman finally stopped the treadmill and began toweling off. He needed the run today. His drill coordinator job had so far been a humbling experience. Kinman considered himself a knowledgeable officer and a pretty decent planner. But planning all these tests for the crew every day had been a strain. Volonino demanded realistic drills, and he wanted them run smoothly. Not all of them had been, and when they weren't, part of Kinman's butt got chewed off.

"Damn, this is going to be a long patrol," he muttered to himself as he walked back to his stateroom.

As Kinman walked to the front of the sub, Marc Churchwell strolled to the back, along the port passageway on the top level of the missile compartment. Churchwell was a master chief and part of the Squadron 16 inspection team that was checking out the *Nebraska*. He was muscular and had a crew cut, a red mustache, and freckles. The month before, he had been the chief of the boat for the USS *Kentucky,* which the *Nebraska* had beaten by just a few points in last year's competition for the Battle-E award. Churchwell hadn't forgotten that. He had volunteered for this inspection trip to see how the competition had bested his old boat.

Well, after a couple days of nosing around, he'd decided these guys weren't gods. In fact, he was a little bit surprised at how ragged their performance had been at the beginning of the patrol, no doubt because they had a new executive officer, along with a lot of new people in the lower ranks. Even so, he had to admit that they did seem to be learning fast. And little things caught his eye that showed the *Nebraska* was a classier boat than others.

Every Trident is built the same, but a crew can add extras to their ship to make it stand out. In the *Nebraska*'s case it was things like a lot of plastic signs hung on doors to identify the offices inside along with operational security warnings posted everywhere. A bulletin board on the activity space next to the crew's mess announced what was playing on the four channels of "WNEB," the sub's compact-disc entertainment system. A small box with a channel selector hung in each sailor's bunk. He could plug his headphones in to listen to country music, rap, or hard rock. Other bulletin boards hung on bulkheads with pictures of sailors who had won awards. Morale also seemed surprisingly high, considering the grueling refit the crew had endured. The sailors appeared energetic with the grunge chores like cleaning, and the chiefs were aggressive.

Churchwell was pleased. He would have been angry if the *Kentucky* had lost the Battle-E to a sloppy boat.

Walking toward the machinery section at the back of

the missile compartment's upper level, Churchwell suddenly stopped. He turned his head to the right, to the jungle of pipes and valves along the hull side of the passageway.

It was a noise. He heard something that didn't sound right.

Churchwell backed up several paces, as if retracing his steps, his ears straining to pick up once more what he thought he had just heard.

He stopped again.

There it was. He began carefully inspecting the pipes and valves in front of him on the hull-side bulkhead. It was a whiny vibrating sound, very slight, but a sound his ears definitely heard. Kind of like "harmonic chatter," he would explain later. And it was a sound that shouldn't be coming from that bulkhead. After twenty-three years on submarines, Churchwell's ears had become acutely attuned to strange noises on the sub. All the older chiefs were like that. They could be sleeping in their beds and wake up just from the change in sounds of different fans on the sub starting up.

Noise could be an insidious enemy on a submarine. The Navy had spent decades making its Tridents the most silent war machines in the world so the enemy couldn't hear and find them underwater. Some of the sophisticated quieting measures were highly classified. The sub's propeller was specially designed to create as little noise as possible when it rotated. Computers in the sub's data processing room monitored the sounds machines on board put out and alerted operators if the equipment suddenly became noisier and needed oiling or repair. The sonar shack also kept detailed records on what sounds the *Nebraska* normally sent out to the ocean, based on readings from its own listening devices. If sonar's ears suddenly heard a new sound coming from the sub, the shack's technicians would scour the boat to find what change in the *Nebraska*'s operation had caused it.

Other silencing measures involved just using common sense. Having all four of the *Nebraska*'s decks hung from the hull, on rubber fittings that looked like giant washers, cut down on the sound from the inside that could escape to the outside. The ocean produces a cacophony of noises on its own from the sea life and currents, but one sound is never heard naturally underwater: metal hitting metal. So acoustical insulation was placed anywhere two pieces of metal touched in the sub. Thick rubber mounts that would absorb sound surrounded machines with metal rotors and gears. Pipe hangers had rubber sleeves. Rubber mounts were attached to every metal valve, to every pipe joint, connection, or bracket.

The crew had to learn to work silently. The clanging from a wrench dropped into a bilge could be picked up by enemy sonar miles away. Signs hung everywhere warning, TRANSIENTS KILL. (A "transient" is a man-made noise.) Rubber strips covered all doorjambs, and every door to every room had a sign on it warning: DO NOT SLAM THE DOOR! THINK QUIET. The loudest noises that more often gave a sub away came from shutting the heavy, circular watertight hatches on board or banging toilet-seat lids. Sailors learned to close the giant metal doors slowly. Toilets were mounted to the deck on tension plates with rubber pads to absorb some of the sound of a lid plopping. Signs were also posted above the johns warning crewmen to put the seat down gently after they finished.

The *Nebraska* could operate at lower acoustical levels depending on how covert it wanted to be. Equipment could be turned off. The sailors could be ordered not to lift weights. Moving the planes made noise, so turns and depth changes could be limited. If it had to, the *Nebraska* could go "ultraquiet" with most crewmen sent to their beds so the sub became a silent ghost ship. The older chiefs remembered lying in their racks for as long as three weeks on some covert operations during the Cold War. Ultraquiet operations were only occasionally needed now.

Churchwell began feeling different pipes with his hands. He finally found it. A check valve from the carbon dioxide scrubber's discharge pipe that sent CO_2 out to the ocean was vibrating. And that was one valve the crew wanted quiet. The noise it was now making traveled up the thin pipe to a hull flange and then out into the ocean. A skilled sonar operator in an enemy sub might be able to sift out that sound and identify the Trident.

Churchwell found a phone nearby and called the control center to report the noise. Control alerted the sonar shack to have the hydrophones on the *Nebraska*'s hull begin listening for any strange sounds that might be coming from the sub. Within minutes, an A-ganger holding a stethoscope walked briskly down the passageway. He plugged the earpieces into his ears and placed the listening end of the stethoscope on different spots along the pipe near the valve, like a doctor checking a patient's chest.

The officer of the deck in the control center ordered the planesmen to begin making a series of U-turns to the left so the sonar array being towed by a long cable behind the sub would drift past the left side of the *Nebraska*'s hull and listen. If the sonar array could hear the vibrating noise from the hull, there was a good chance an enemy's sonar could as well. Even if the array didn't pick up a noise, repairmen planned to fix the check valve so it no longer vibrated. Technicians would also write a detailed report describing the noise they had heard from the valve and pass the document on to other subs in the fleet. That way, if the *Nebraska* or any other Trident heard that faint whine again, the crew could quickly recognize the sound and pinpoint its source.

The vibrating check valve was an easy fix. The next headache Volonino faced that day wasn't. The long cable that towed the sonar array behind the sub was stuck. The cable had snagged on the roller it wrapped around inside the sub and the hydraulic system couldn't reel it in. Volonino couldn't sail back into port with a two-thousand-foot cable trailing the sub in order to have it fixed. The last thing he wanted was to have to cut the cable. It would leave him with only one other towed sonar array for the patrol. But more important, the towed array cost more than $1 million. The Navy didn't have a lot of replacements and didn't take kindly to sub captains who left one of them at the bottom of the ocean. Foul up this repair and he could be writing a check paying for the array himself. News of the stuck cable spread like wildfire through the sub, so that by the time Volonino announced it on the speaker system most of the crew already knew he had a big problem on his hands.

After an hour and a half of tinkering with equipment inside the sub failed to unlock the cable reel, Volonino decided to surface the boat. Repairmen would have to climb inside the superstructure encasing the top of the vessel to reach where the rollers that spooled out the two sonar arrays sat near the stern, and then see if they could unsnag whatever had tangled up one of the cables.

It was cramped, dirty, wet, and dangerous work. Even with the sub surfaced, water sloshed around the free-flooding area where the reel was located. Any wrong move and a sailor could lose a hand or finger as the taut cable shifted on the roller. Fortunately, the seas were calm outside, so the men crammed into the watery compartment wouldn't be bounced around too much while they worked. But no one was looking forward to the chore.

As the sub reached periscope depth, a half-dozen sailors picked for the repair crew gathered at the top level near the sub's stern and began climbing into orange canvas harnesses and strapping life vests around their chests. When the *Nebraska* surfaced, they would climb up the ladder, open the hatch above, and crawl out onto the top of the deck, hooking harness lines to runners so they didn't fall over the side. Then they would lower themselves down into the superstructure near the back of the Trident where the cable reel sat.

But before they did any of that, the sailors decided to have some fun. One of them summoned Beasley to the back of the boat.

Submariners are notorious for playing practical jokes on one another and the *Nebraska* crew was no exception. Volonino was off-limits; he took only so much kidding. But everyone else was fair game. Tell a shipmate you're looking for another crewman and he'll likely tell you he saw him at the other end of the sub just to see if you'll make the long trek to find him. For the past week, Kinman had been slipping a heavy brass

plaque into Dave Bush's bed every night. Every time
Bush hopped into his rack, he kept banging his knees
into it. Duane Ashton, the executive officer on the last
patrol, had the door to his private stateroom stolen. After several days of sleeping and dressing for all the
world to see, he finally posted two junior officers at his
entrance as guards and angrily told the wardroom that
they would stay there until the door was returned. Big
mistake. Never let a joke get under your skin. It just invited more pranks. Kinman and Hardee, who had
stolen the door and hidden it in the missile compartment, screwed it back onto the stateroom entrance. But
a month and a half later, two other crewmen stole it
again.

Newcomers were a favorite target. Nubs were often
told to grab slices of bread from the galley and go to
the engine room to feed the "shaft seal." When they
walked back to the stern, a chortling technician told
them, no, the *Nebraska* had no animals on board. Then
he showed them the seal that wrapped around the shaft
extending out of the hull to turn the propeller.

Beasley was now about to be the victim of one of the
crew's favorite jokes. As soon as he arrived at the upper level where the other repairmen were dressing out,
Jose Victoria, a petty officer first class in the missile
section, threw him a harness and life jacket and ordered Beasley to quickly put them on. Then Victoria
handed him a cranial helmet like the ones aircraft carrier deckhands wore, plus large blue headphones to
protect ears from loud noises, and a pair of goggles.
Beasley put them all on as well. He looked like a space
cadet.

Victoria next shoved a ten-foot-long wooden boat
hook into Beasley's hands and looked sternly into his

eyes—or, to be more accurate, through the goggles and into his eyes.

"Beasley, you're being given the most important job on this sub right now," Victoria told him gravely. The other petty officers suiting up turned their backs so Beasley couldn't see them snickering. "When the sub surfaces, I'm sending you topside to retrieve the mail buoy."

"The what?" Beasley asked.

"The mail buoy," Victoria repeated. "There are a lot of men on this boat who've been waiting desperately for the mail to come. And they're going to be awfully pissed if they don't get their letters from home." Beasley's mission: climb up the ladder when the boat surfaces and, once he's on top, attach the tether from his harness to one of the deck clips, then lean over the side of the sub and snare the buoy carrying all the mail for the *Nebraska* with the boat hook. "If you miss that mail buoy, you get disqualified for the rest of the patrol."

"Huh?"

"Now pay attention!" Victoria shouted, shaking Beasley by the shoulders. "It's dangerous up there. You're going to kill yourself if you don't watch out."

Now Beasley was worried. He began sweating. The lenses on his goggles started steaming up.

"You got any wipettes for this thing?" he said, terrified he wouldn't be able to see what he was doing topside. "This thing is fogging up."

Victoria handed him a rag. Beasley lifted the goggles off his eyes and struggled to wipe the inside of the lenses. He began pacing back and forth as the other men gathered the tools and gear they would need topside. Beasley was so fidgety and nervous he never stopped to think just how the hell the Navy would ever put a buoy with the *Nebraska*'s mail in the middle of the Atlantic and expect the sub to find it.

"How far over am I going to have to lean to hook the buoy?" Beasley asked frantically.

"Don't worry, I'll show you when we get up there," Victoria answered. "This is on-the-job training."

The goggles continued to steam up. "This isn't going to work," Beasley said, panicky. "I hate this fricking helmet. . . . Is this like a standard buoy or is it moving?"

"It's moving," said Katherman, who was dressing out next to him.

"How do you know where to get it?" Beasley finally began to wonder.

No one answered.

Finally, Chief Spooner walked over to a phone, picked it up, and pretended to be listening.

"The mail buoy has been delayed," he said, hanging up. "We won't pick it up until tomorrow."

Beasley looked as if he'd gotten a reprieve from death row. "Should I leave this stuff out until tomorrow?" he asked.

"Naw," the petty officers around him said in unison.

It was after 10 P.M. when Volonino finally got back to his stateroom. He peeled off his dirty sweat-soaked poopie suit and headed for the private shower between his quarters and Boyd's, which the sub's two senior officers shared.

Volonino's stateroom was cozy and was the best-outfitted on board. It served as a combination bedroom and office. Along the right bulkhead he had a built-in desk with file drawers and a personal computer. On the shelves overhead sat a CD and tape player along with stacks of manuals, reports, and incoming paperwork. To the right of his desk stood a locked cabinet with a combination safe on top, where he kept classified documents and one set of keys used for launching the missiles. On the left bulkhead he had hung a small plastic board for writing reminders and pasted up Polaroid photos of crewmen to whom he had presented awards. His bed folded out at the far-end bulkhead and had a red University of Nebraska blanket draped over it. Above the foot of the bed hung a box with constant digital readouts of the sub's depth, course, and speed. A speaker box built into the bulkhead next to his desk was connected to an open mike in the control center one floor above so Volonino could eavesdrop on the conversations of the officers and sailors driving the sub.

Though it was the only private room in the sub besides Boyd's, Volonino found he really had no privacy. The sound-powered phone near his desk constantly whinnied like a horse when someone called. Many a young sailor had barged in with a report and caught the commander toweling off naked from a shower. Or there was the terrified radioman who had to wake him up in the middle of the night to read a message. Sometimes the kid would just stand there silent hoping the CO would wake up on his own, and Volonino would bolt up terrified because he could sense someone was in the dark room just breathing.

Volonino finished showering and put on a clean poopie suit, hoping it would revive him some. It had taken five hours to fix the towed sonar array so it could be reeled back in. Weller had recruited some of the sub's strongest sailors for the strenuous work, and Volonino had crouched inside the superstructure compartment to supervise the work personally the entire time. Normally the captain stayed at the bridge, but this repair was dangerous, so Volonino had decided to make sure it was performed safely and to let Boyd take care of driving the boat.

The cable that was wrapped around one of the rollers used to spool out the sonar array smoothly had slipped from its track and kinked. The repairmen had to take apart the roller in order to work out the kink so the inch-thick wire could be reeled back in. It had been a sunny, ninety-degree day in the Atlantic with the boat rocking side to side gently and a warm Gulf Stream breeze blowing across the deck. But inside the super-structure, it was hot and wet, and as the sun set it became darker and scarier in there.

The crewmen had begun muttering to themselves that the *Nebraska* seemed to be cursed with mechanical problems. But Volonino was still upbeat. So far, he privately rated the crew's performance average to above average. There were always equipment glitches the first week of a patrol. Some, like the noisy check valve, were easy to fix, while others, like the tangled cable, were a bear. Rarely did Volonino have to fix the same problem twice. He didn't know if that was a blessing or a curse. The sub was so complex, one of a million things could break at any time. That sure made the job lively.

Volonino slumped into the chair in front of his desk. He still had a stack of evening reports to read, then Hunnicutt, who seemed to be a night owl, had invited him to a late poker game in the wardroom. Volonino couldn't refuse. But he didn't know if he'd be awake enough to even focus on the cards.

13 • Sunday

Habermas draped a yellow covering over the wardroom's dining table, then carefully spread out a gold-fringed red silk altar cloth at the head of the table. From a large black box he pulled out red missals and yellow booklets for a Catholic Lay Eucharistic Service and placed them in front of chairs on each side of the table. At the front of the table, normally reserved for Volonino when the officers ate, Habermas sat down and reached into his box again, this time for a silver crucifix on a stand, which he placed on the red mat to his right, and then a white candle in a silver candleholder, which he placed on the mat to his left.

Habermas had several minutes before the 7A.M. Sunday service was supposed to start, so he began reading to himself from a prayer book. Chaplains did not ride on the Tridents. Instead, a crewman was appointed as a lay leader to conduct a Sunday service. This was the third patrol that Habermas had been the *Nebraska*'s Catholic lay leader. The base's chaplain had interviewed him for the job—mainly to make sure that he was a practicing Catholic in good standing—then Habermas and the lay leaders for other subs had attended a one-day retreat back at Kings Bay to learn how to conduct the ritual. Several days before this patrol, he had visited the church on base to have the chaplain bless him and consecrate his Communion wafers.

Habermas had tried to be a good and moral Christian. He didn't consider himself deeply religious and he knew he wasn't perfect. He had his flaws, like any other human being. But he believed deeply in God and Jesus Christ, and he went with his wife, Sally, to Mass every Sunday when he wasn't at sea. And when he was on patrol, he found that conducting this service once a week helped him to deal better with his inadequacies.

This Sunday service would be a lonely one for him. It usually was the first Sunday on patrol. Sunday was the crew's one day to sleep in and the men were just too worn out after the first week to get up at 7 A.M. for church. Volonino, a Catholic, usually attended Habermas's service. But after a week of less than four hours sleep each night, he was exhausted and desperately needed to recharge. Habermas understood. He didn't take it personally. Many of the men attended Mass regularly when in port, but at sea the work schedule was brutal, so they tried not to miss any chance for eight hours of uninterrupted sleep. On future Sundays, Habermas usually had two to four men show up. About a dozen regularly attended the Protestant service.

Habermas lit the candle with a Zippo lighter and began the service. Even when no one showed up, he always read it silently to himself. It helped him recharge his own batteries. He conducted an abbreviated version of the ritual, cutting short the homily and summarizing from the Sunday missal. If he had a congregation, he would discuss a scripture reading with the men who attended. Then he delivered the Profession of Faith, offered Communion, and finished with a Sign of Peace and a concluding prayer.

"Lord, I ask that You bless all the crew members of the *Nebraska*," Habermas finally said. He blew out the candle, then packed the silver pieces, the books, the pamphlets, and the altar cloth into his black box.

Rodney Mackey walked into the wardroom several minutes after Habermas had left. A twenty-nine-year-old petty officer first class and one of the sub's electricians, Mackey led the Protestant service. (Mackey and Habermas knew of no Jews or Muslims on board.) Nine years ago, he had been born again. Until then, he had not known much about God, had not realized that he needed to be saved. Then aboard an attack submarine he had attended a Sunday service, and the officer who led it asked each man in the congregation, "If you died today, where would you go?" Mackey danced around the question because he didn't know if it would be heaven or hell. Three weeks later when the sub returned to port, Mackey considered himself saved. An African-American, he became a deacon at the Citadel of Hope Church of God in his hometown of Leesburg, Florida. Sure, over the nine years there had been stumbles along the way. But God always helped him to pick himself back up.

He had applied for the job of Protestant lay leader when it opened up on the *Nebraska*. Mackey had no religious training, save for the one-day retreat with the base chaplain. But he had no doubt that God was using him to spread His message.

Mackey placed a CD with Christian folk music into the wardroom's disc player and hung a white bulletin board on the aft bulkhead. As the folk music began, he wrote "GOD," "Lukewarm," and "SATAN" on the board. Mackey had a little better luck than Habermas for the first Sunday. A half-dozen sleepy-eyed Protestants walked in at 8 A.M.

Mackey's service blended old-style revival with modern folk music. As the men settled into their chairs, he passed out sheets of music and punched the channel selector on the disc player to another song.

"I want you to think about what you're singing," Mackey said with his eyes closed as the music began.

The men mumbled the verses.

> *Lord, prepare me to be a sanctuary*
> *pure and holy, tried and true.*
> *With thanksgiving, I'll be a living sanctuary for you.*

Mackey then read a passage from I Corinthians. "We are all one body, one crew," he said, closing his Bible. "We all work together. In the body of Christ may you be strong."

Mackey turned toward to his board. "Have you turned to God, or to Satan, or are you lukewarm?" he asked the men. They must shun Satan and refuse to be lukewarm, Mackey implored in his sermon. "Stay in position with God," he said. "You never know how many people are following you."

Mackey had a crewman read a verse from Revelation. "The thought for the day," he said when the sailor finished, "is true repentance turns from the wrong and returns to the right."

They sang another religious folk song, then Mackey recited a concluding prayer. "Since the last patrol, have you turned away from God?" he finally asked the men. "Stay focused. Stay focused on God."

The service concluded, Mackey stood at the wardroom door and shook each man's hand, saying softly to each one: "Stay in position. Stay in position with God."

When the last sailor had left, Mackey ejected his CD from the disc player and unhooked the bulletin board from the bulkhead. Then he gathered up the song sheets the men had left behind.

He had just enough time to return to the locker in the berthing compartment where he stored his religious material before he was due in the control center.

Their worship over, Rodney Mackey, Steve Habermas, and the other men of the *Nebraska* would now practice how to destroy much of what God had created.

14 • Cookies and Cream Them

The eight hours of uninterrupted slumber worked the miracles it always did. Volonino's head no longer had that constant dull throb from too many nights with too little sleep. His eyes no longer felt scratchy, his limbs no longer ached. He seemed a new man as he reached up for the overhead microphone at the conn.

"Russia is in economic chaos," he announced on the sub's speaker system. "It's out of oil, out of food, and its missiles are on alert. U.S. strategic forces have therefore been put on alert." Volonino felt obliged to provide a scene-setter for these missile drills. The brass liked that. To start off strategic exercises back on shore, the admirals always gave the Trident commanders elaborate political scenarios with the world going to pot and nuclear weapons having to be used. But Volonino always found the scenarios unrealistic.

The Defense Department almost always generated the simulated launch orders for training exercises like the one Volonino was about to conduct. Surprisingly, American Presidents rarely participate in war games involving the release of nuclear weapons. Shortly after they're inaugurated, all of them receive a Pentagon briefing on targeting options the generals could offer with nuclear weapons and on the officer who would be constantly at a President's side with the "football." That is the nickname for the briefcase the officer carries with attack plans and the codes for releasing nuclear weapons. But after that, most Presidents prefer to avoid the grisly subject. When the Pentagon holds a high-level war game with nukes, most send an aide to play the role of commander in chief.

"Set Opsec, condition Alpha," he said over the microphone, then hung it up and returned to his stateroom one level below.

Operational Security Alpha was one of the quietest operating modes the *Nebraska* had. Few water spigots could be turned on. Most fans were turned off. The electric cords for the washing machines were pulled from their plugs. The trash compactor was stopped. The weight lifting set was stowed away.

One of the planesmen wiped off the grease-pencil markings on the sign posted over the ship's control station, then wrote in the new defense condition under which the sub now sailed. It had skyrocketed, from Defcon-5 (the normal peacetime level) to Defcon-2 (the second-highest alert level).

Kinman became fidgety. In several minutes he had to replace Habermas as officer of the deck, and, with all the drills he had been coordinating lately, he hadn't had time to brush up the procedures he must now supervise at the conn. He didn't want to bungle this watch, not when the sub was about to practice blowing up the world.

Kinman hurried back to the navigation room aft of the control center for a quick check of the sub's position in the Atlantic. He walked quickly through the control center again, this time past the forward entrance, and popped his head into the sonar shack.

"You got any contacts?" he asked in a rush. Sonar had none.

Kinman walked back to the conn, reached into a shelf under the railing around it, and pulled out a secret Opsec Alpha manual that listed all the machines that had to be turned off. He began comparing the list with a diagram of the ship he had unfolded.

"Are we all rigged yet for Opsec Alpha?" he asked Habermas.

"We're in the process of doing that," Habermas said calmly.

Using a grease pencil, Kinman made a show of checking items on another loose-leaf binder for the officer of the deck that contained the steps he had to accomplish before the missile drill began. Habermas gave him a last-minute cram course of do's and don'ts, while Kinman pounded his forehead with his fingers, as if he were nailing facts to his brain.

"Officer of the deck, the ship is rigged for Opsec Alpha," announced Sean McCue, who was serving as the watch chief at the ballast control panel.

"Very well, chief of the watch," Habermas said to McCue, then turned his duty over to Kinman.

"Attention helm and quartermaster," Kinman announced loudly to the planesmen and navigators in the room. "Lieutenant Kinman has the deck and the conn."

Still nervous, Kinman hopped down from the conn and walked to McCue's seat in front of the ballast control panel to make sure water was distributed properly in the ballast so the sub would stay level for the missile launch. The *Nebraska* stayed underwater to fire off its missiles, but at a shallow depth.

Keeping the vessel stationary at that depth as the missiles shot up was difficult, however. The force of the launch pushed it down. Then the sub bounced up after each launch because it suddenly weighed 130,000 pounds lighter with one missile less. The crew let seawater rush into the empty tube to help compensate, but that water was lighter than the missile. So to keep the sub from bobbing out of the proper launch depth, computers took control of pumping tons of water in and out of special ballast tanks called missile compensation tanks, which all had to be done faster than McCue could flip switches at the control panel.

Kinman began leafing through more notebooks to piece together a torpedo evasion plan after the launch. Once a Trident fired one of its missiles, the rest of the world would know its exact location. Navy attack subs in the area could protect it only so much. After a launch, a Trident had to hide from what would be a very angry enemy, whose sonar shack suddenly had a blast of valuable noise for closing in on the kill.

Kinman could play possum. An enemy torpedo would home in only on a target that was moving. Otherwise, the torpedo would strike mountains in the ocean. Kinman might have a better chance of living if he kept the *Nebraska* standing perfectly still after a launch so an enemy torpedo would mistake it for a rock and speed by. But it took guts not to run, and Volonino would have to make that call, not he. Kinman decided his evasion plan would be to kick the engines into full throttle and sink the *Nebraska* as deep as he could to get out of there.

"It's about to get hot," Kinman murmured under his breath. These graduate-level war games were intense.

In the radio shack forward of the control center,

alarms began beeping on the standard information display consoles as the emergency action messages (EAMs) began interrupting routine radio traffic and flashing on the screen. The radio operators at the two SIDs began calling out the first three or four highlighted lines that scrolled down on their screens, which alerted the shack to the EAMs. Liebrich quickly opened the safe that stored the cryptographic manuals and code books they would need to decipher the EAMs that had begun pouring in.

The Pentagon assumes that in a nuclear crisis its communication centers would come under enemy attack, so it has dozens of ways to send messages to Tridents over scores of independent transmitters based in space, on land, at sea, and from airplanes. When the Strategic Command sends out an emergency action message, it is automatically relayed to all these stations for transmitting to the Tridents. The *Nebraska* wasn't getting just one EAM. Dozens of them were pouring into the sub. STRATCOM was bombarding the Trident with the same message.

Liebrich ripped pages off the printer and began decoding the first six groups of scrambled letters. Dan Montgomery, the radio shack's chief, had already arrived and was watching over Liebrich's shoulder as he decoded. It was an EAM for the *Nebraska*. Montgomery had handled so many emergency action messages in his twelve years in the Navy that the chore had become routine. The only time it now excited him was when his shack was tested on how fast it could process them. Montgomery didn't know what his reaction would be if the consoles flashed a real order to launch. He prayed he'd never have to find out.

Montgomery rushed out of the radio shack and to

the conn, where Kinman paced back and forth. The chief spoke in a low voice to the lieutenant, then Kinman reached up for the microphone over the conn.

"Alert one, alert one!" Kinman announced on the sub's speaker system, his heart now thumping.

Officers throughout the sub who weren't standing watch dropped paperwork on their desks, swallowed one last mouthful of lunch, or rolled out of racks to dress quickly. They all sprinted to the control center. Six of them were assigned to decode and process an emergency action message. Volonino had divided them into three two-man teams. When an alert one was announced, the first two-man team to reach the radio shack began processing the message.

Thorson and Joe Davis, the *Nebraska*'s supply officer, were the first to arrive at a door just to the right of the control center's forward entrance. A small sign posted on the door warned menacingly:

Commanding Officers
Operation Control Room
RESTRICTED AREA
Authorized Personnel Only

The door had a peephole, and a combination lock above the knob. Montgomery opened it and Thorson and Davis quietly slipped in. Rob Hill and Bob Tremayne, two lieutenants junior grade on board, were the second team to arrive and walked in behind Thorson and Davis.

The small op-conn room, which was connected to the radio shack by another door forward, was now cramped with the four officers and Montgomery inside it. Fake wood paneling covered the bulkheads. Locked

cabinets and safes arranged on the port side stored
code manuals and cryptographic material. On the starboard side, a bed could fold out. The op-conn room
was originally envisioned to be where the sub captain
could sleep during wartime so he would be close to the
radio shack and the messages that came in. But
Volonino, like most sub captains, preferred to remain
in his more comfortable stateroom, which was less
than twenty feet away on the second level, so the op-
conn room was being converted to a computer work
center for the radio operators. But it was still used to
decode incoming EAMs.

The four officers spread out code books on a table,
which could be folded down from the starboard bulkhead instead of the bed. They began unscrambling the
rest of the message, reading off the four-letter groups
and comparing them to the decryption instructions in
the manuals. The decoding had to be done by hand, but
the men worked quickly. Most EAMs took less than
ten minutes to decode if the message was intact.

This one wasn't. In wartime, any number of things
could garble a message: enemy jamming or attacks on
transmitters, rough seas or atmospheric disturbances
that broke up the reception. The last thing the Navy
wants is a Trident captain confused about his orders, so
the *Nebraska* had thick, top-secret manuals with elaborate instructions and contingency plans along with
page after page of options matrices for how its officers
should deal with garbles. During missile drills, the
Strategic Command rarely sent a clean message. Most
had missing information or code letters mixed up to
test how quickly the crew could resolve errors. With
dozens of the same EAM coming into the radio shack,
the officers could often piece together a readable mes-

sage from the passages in each transmission that were intact. The manuals also spelled out in excruciating detail "garble criteria"—how many and what kind of typographic errors the officers could accept and still deem the message valid.

After cutting and pasting, the four officers in the opconn room had produced a readable message. This one wasn't yet an order to launch. There were a number of "informational" EAMs that would stream into a Trident before the President finally ordered a missile launch: secret intelligence on enemy ships, updates on world turmoil, changes in the Defcon for U.S. strategic forces. This was one of them. Thorson and Davis grabbed the final version and carried it to the conn.

Volonino and Boyd stood on the conn, quietly leafing through manuals and documents with orange "Top Secret" covers over them. A large brown lawyer's briefcase sat between the two men; it was packed with the classified documents they needed for the launch. The CO and his executive officer had been busy reviewing dozens of checklists, conferring with Kinman on the quieting measures and evasion plan, and receiving briefings from Fred Freeland, their weapons officer, on the status of the sub's missiles and torpedoes.

The control center became crowded now, with *Nebraska* crewmen involved in the launch and squadron riders watching with notepads open to make sure procedures were followed to the letter. The hushed conversations began to pick up. Hunnicutt stood near the rear entrance to control on the port side. "I wonder if they have *Dr. Strangelove* on board?" he asked quietly with barely a smile on his face. "That's always a good movie to watch before and after a missile drill."

The radio shack had also sent to Volonino and Boyd

separate copies of the EAM at the same time the four
men in the op-conn room reviewed it. This way the
skipper and his XO could independently decode and
piece together the message. When Thorson and Davis
arrived at the conn, all four of them had read the message. All four now had to agree on what the message
meant. If any one of them had a question, stacks of
classified manuals were nearby with instructions on
how to resolve discrepancies.

The launching of nuclear weapons is such a monumental act for any human being to take that the U.S.
military's high command has left nothing in the procedure to chance. Everything the men now did, every action they took with the missiles, every word they
uttered, was scripted. Redundancy was built into every
decision they made, no matter how minor it was. Nothing was trusted to memory. There were safes on the
sub bulging with confidential manuals with procedures
written out for everything the missile technicians could
do with the nuclear weapons. The officers had practiced receiving EAMs and simulating missile launches
so many times it came to them as naturally as tying
their shoes. But for every missile drill, they still hauled
out thick manuals to verify the tasks they had to perform. A technician couldn't remove the cover from a
missile compartment equipment cabinet without reading a checklist on how to unscrew the bolts.

Thorson and Boyd stood at attention about a foot
from the lip of the conn, both of them holding the single piece of paper in front of them, as if they were
church ushers about to deliver a collection plate to the
altar. "Message number one on the conn," Thorson began his lines.

Boyd, who served now as the traffic cop for EAMs,

motioned the two men to approach. "Report message one," he ordered, giving Thorson permission to speak.

"Report message one, aye, sir," Thorson repeated. "Captain, message one is a valid informational message. Piecing was required. Authentication is not required."

Emergency action messages that called for launching the missiles required a special authentication procedure to verify that the sub was receiving a valid order. EAMs that just provided information didn't need the extra safeguard.

"I concur," Davis said softly.

"I concur," Boyd repeated. "Authentication is not required."

Volonino also agreed. "Very well, what are the instructions?" he asked.

Thorson read them. The other three men concurred on what they meant. The world was becoming more tense. The Defcon level soon would be raised to 1, the highest alert level.

Volonino sat back in his captain's chair and began quietly reviewing top-secret manuals as sailors, chiefs, and officers busied themselves around him like bees in a hive. Hunnicutt, standing to his left and sipping coffee, became philosophical. He had never gotten used to these exercises. "Not when you're blowing up half the planet," he said gravely. "We're talking about the kind of warfare that's beyond anyone's comprehension but God's."

Volonino was convinced that they must practice the missile launches with the same intensity they had during the Cold War. The genie could not be put back in the bottle. Nuclear weapons couldn't be uninvented. It was not enough simply to have the Tridents on patrol, he believed. For deterrence to be credible even today, the subs had to maintain their finely tuned capability to fire missiles at a moment's notice. The missiles sitting idle in the tubes had to be electronically monitored to ensure their reliability. The crews had to constantly run launch drills.

Just as important, other nuclear weapons states had to be convinced that the United States could launch the missiles at any time. Every year, two Tridents—one from the Atlantic Fleet, the other from the Pacific—were picked at random. Four missiles with dummy warheads were loaded into them and fired at points in the ocean. Tubes in the *Nebraska* from which test missiles had been fired had painted on them the sub's emblem and the date of the shot to commemorate it. Tube five, for example, bore a medallion commemorating the submarine's first test launch, on August 20, 1993. Russia was notified in advance of the test launches, not just to calm jitters but also to demonstrate to the Kremlin the confidence the United States still had in its strategic arsenal. So far, every test launch had been successful. They had to be. Volonino shuddered at what Russia would think of America's nuclear credibility if there were launch failures.

Volonino knew he'd have to be brain-dead not to realize the enormity of what he was now rehearsing. The hair on the back of his neck stood up every time the sub pretended to go to nuclear war. It was almost impossible for him to comprehend the power his country had placed in his hands. The responsibility would have been just as grave if they had given him only one missile with one warhead that could kill one million people. Twenty-four missiles with more than one hundred nuclear warheads? Who could ever be emotionally prepared to unleash that kind of megatonnage?

Volonino had thought a lot about why he did this kind of work. And like most submariners, he kept his feelings to himself. Patriotism motivated him the most. There weren't fifty people out there in the civilian world lusting for his job. But someone had to sit under this ocean for months at a time. Someone had to make the sacrifice, someone had to be ready and willing to launch the most destructive weapons ever created if that was what it took to preserve American ideals. Volonino thought his men felt the same way. They were all proud they were doing a job few others wanted, he believed.

Volonino had no doubt that he would launch the missiles if the President of the United States gave the order. He was sure the threat to America would have reached the point where a President had no choice but to issue the order.

But what if the political leadership wasn't responsible? What if he was being ordered to launch because of the mistakes of politicians? What if his politics were the opposite of a President's?

It didn't matter, Volonino fervently believed. A Trident commander's politics didn't matter. Dating back to colonial days, the American military had a tradition of being submissive to civilian control and direction. Volonino had sworn a solemn oath when he was commissioned to obey orders from his President. He took that oath seriously. He might not like what a President said on TV. But he wouldn't hesitate for a moment if his commander in chief told him to launch.

Neither would his officers. During simulated launches, most of them had little trouble identifying the cities from the target coordinates. The first time they did, it was scary realizing that the row of numbers represented millions of human beings who would perish. After hundreds of drills, the anxiety dulled. Occasionally a target package that came from STRATCOM would pique their interest, particularly if it included a hostile country just in the news. Otherwise, practicing to launch missiles became just another ship procedure they performed over and over again, like diving and surfacing the sub. They were all technicians, paid not to judge its correctness but rather to make 100 percent certain that the missiles could be fired if the President ordered it. None of the officers disagreed on that point.

The American bishops in Volonino's Catholic Church had practically condemned any use of nuclear weapons as immoral. But Volonino was more worried about what God would think of him than what the bishops did. Every service member faces this dilemma, whether he is a lowly infantry private or a Trident sub captain. The military is in the business of taking lives.

What would God think of him if he had to launch his missiles? Volonino didn't consider himself an expert on the Bible, but he was sure there were scriptures that answered the question. If your cause was just and noble, if you followed God's teachings and believed in them, he was sure the Lord would understand.

Several more informational EAMs arrived. Thorson and Davis or Hill and Tremayne marched up to the conn each time and presented them. In a real crisis, there could be weeks of informational EAMs streaming in as tensions rose. That time was compressed into several high-stress hours for this drill. Volonino decided to sneak off to his stateroom to catch up on paperwork until the next message.

1:35 P.M.

He returned to the conn ten minutes later. The radio shack had announced over the speaker system that it had another EAM. Levering followed close behind with a soda for him in a plastic cup.

"Ah, can't fight a war without Diet Coke," Volonino quipped, then thought about the further humor in it. "Seems illogical to worry about your weight if you're going to blow up the world."

Boyd handed him a copy of the message he had already begun decoding.

In the op-conn room, Thorson and Davis were busy decoding as well. Of course, the message was garbled, so they had to splice again to come up with a complete message. But this one, they quickly discovered, was not informational. In a real war, it was one of the most highly classified orders the military would ever send. It was an order to launch.

Calling out four-letter groups, Thorson and Davis wasted no time decoding the message. Nuclear control orders, as they were called, always had the same format, with the same four parts. The first part of the message told the *Nebraska* the war plan STRATCOM wanted the sub to execute—in other words, how many nuclear weapons it wanted launched and the coordinates of the targets they were supposed to hit. The Pentagon's strategic war plans were flexible and constantly being updated. The President could strike practically any target or combination of targets he wanted. Every conceivable option the generals could dream up was on his menu, from a single surgical strike, to limited attacks, to all-out nuclear war. Volonino could empty his missile compartment. He could launch several missiles or only one. Or if the White House wanted just a single nuclear detonation, he could shoot a D-5 missile with all the warheads atop it electronically dudded except for one. Any type of holocaust Washington wanted, Volonino could provide it.

The second part of the message spelled out the date and time window the Pentagon wanted the missiles fired. The strategic war plans were carefully choreographed. Exact times were prescribed for atomic bombs falling on targets. The war planners didn't want the *Nebraska*'s warheads exploding over an area outside the window and frying Air Force B-2 bombers swooping in at the same time to drop their nuclear payload.

The last two parts of the message assured both Volonino and the President of the United States that the two men could trust each other. They contained the combination to a safe and the codes for the cookies.

The combination was a recent innovation. The old refrain that the two most powerful men in the world were the President of the United States and a Trident submarine captain always sounds dramatic, but it has never been completely true. The firing of intercontinental ballistic missiles from land could always be controlled by a higher headquarters with electromechanical interlocks. A Trident submarine doesn't come with a five-thousand-mile extension cord. To fire the Navy's nuclear weapons, a radio message has to be sent to a Trident and then that sub, on its own, launches the missiles. Neither the civilian nor the military leadership has ever been completely comfortable with this arrangement, however, so over the years elaborate safeguards have been put in place to keep a rogue submarine from launching missiles without an order from the President.

The Navy built its Tridents so no one person could
fire the missiles by himself, not even a captain. It took
four sets of keys to unlock electronic and mechanical
devices throughout the *Nebraska* before any of its missiles could leave their tubes. Volonino kept the first set
in the safe in his stateroom: twenty-four firing unit
keys strung on two-foot-long olive-drab lanyards. The
keys were inserted into the gas generators attached to
the twenty-four missile tubes to arm them for launch.
Freeland, the *Nebraska*'s weapons officer, had the second key, called the tactical mode key, which he kept in
a safe in the second-level missile control center. Only
he knew the safe's combination. Before a firing, Freeland had to stick that key, painted green, into the missile control center's launcher panel and turn it, in order
to electronically line up the sub's firing computers for
a launch. The third key was actually a trigger, which
Freeland also kept locked in his safe and which he had
to squeeze in the missile control center to blast the
missiles out of their tubes.

The fourth key, a purple one, was called the captain's
indicator panel key, or CIP key. The sub commander inserted it into a large gray metal box in the command
and control center just forward of the conn. The CIP
key was one of the last that had to be turned to complete
electronic circuits and activate the weapons system.

Until the early 1990s, Trident skippers kept the CIP
key in their stateroom safes. But during the Bush administration, Congress became worried that all the
fail-safe measures in place to guard against an accidental nuclear launch were still not enough and ordered
the Pentagon to appoint a panel of outside experts to
review them. That panel, called the Commission on
Fail Safe and Risk Reduction, was chaired by former

United Nations ambassador Jeane Kirkpatrick and was one of the most secretive groups ever assembled by the U.S. government. For the first time, the Pentagon divulged to outsiders some of the most sensitive information in its vaults—how the military controlled the use of nuclear weapons.

The commission, too, feared that the safeguards still weren't airtight, and it ordered changes, which the Pentagon adopted. Among them, the Trident captain would no longer control the CIP key. The Strategic Command would. Expensive new safes were installed in every Trident. In the *Nebraska*'s missile control center, behind where Freeland stood during a launch, sat three heavy safes, painted tan, stacked one on top of the other. On the front of each safe were two combination locks, and inside each safe was stored copies of the CIP key. The safes were guarded in the missile center twenty-four hours a day by two sailors, and an alarm sounded throughout the sub when someone tried to open them.

Also, no one on board had the combination to these safes. Those numbers came in the third part of the emergency action message that Thorson and Davis translated. The final electronic link needed to launch the missile would come from shore.*

The fourth element of the message contained a row of randomly arranged numbers and letters for the

*The Kremlin trusts its submarine crews even less and has tighter controls over Russian nuclear weapons. There are four access doors on each of a Trident's missile tubes, which seamen can open under carefully controlled conditions to repair the rocket's stages. The missiles on Russian subs, however, are hermetically sealed in their tubes with no way for the crew to even touch them. To fire, a Russian sub captain also must first receive a special code in the launch order from higher headquarters, which his crew then types into a computer on board to electronically unlock the missiles for launch.

Sealed Authenticator System code, one of the most closely held secrets in the U.S. government. A Trident has to have some way of being absolutely sure that the launch order radioed to it is legitimate. The crew have to be confident that the emergency action message actually comes from the President, that a hostile country or a rogue American general or simply an impostor hasn't broken into the defense communications network and transmitted a phony order to start World War III. The Sealed Authenticator System code is the final step a Trident would take to verify that the order is for real.

The supersecret National Security Agency, whose spy satellites and ground antennas vacuum phone calls all over the world, produces the SAS codes. Agency machines stamp the same computer-generated code of randomly arranged letters and numbers on two plastic cards. The machine then seals each card in a shiny metal foil. The code cards are nicknamed "SAS cookies" because they look like wafer bars wrapped in tinfoil. The machine was specially built to do all the stamping and sealing itself, so no human eyes ever see the numbers and letters printed on the cards.

One of the sealed cards is placed aboard the Trident. Its twin, with the identical arrangement of numbers and letters, is kept by the Strategic Command. When STRATCOM's generals drafted the emergency action message to launch nuclear weapons, they would break open the sealed card and print its authentication code in the order. At the other end, the Trident captain could break open the card he had and compare the alphanumeric code on it with the arrangement of numbers and letters in the message. If the two codes matched, the captain could be certain that he had a valid launch order.

Thorson and Davis finished deciphering the message. If this had been a real launch, they would have opened a safe above a crypto vault along the room's port bulkhead, which stored the SAS cookies. The sealed authentication codes were delivered to the *Nebraska* under tight security. Even if a spy managed to steal one, it wouldn't do him much good. The cookies are specially manufactured so that if someone unwrapped one to copy its code, the card could not be resealed and put back into the batch. The *Nebraska* kept boxes full of the cards on board. If terrorists hijacked a batch of them on land, the Navy would simply send out a radio message to the sub to use other cards.

The SAS safe actually was two safes in one. The outer door, with a combination lock on it, opened to an inner door with another combination lock. Behind the second door were stacked the cookies. Thorson had the combination for one of the locks. Davis had the combination for the other.

Thorson and Davis took the emergency action message to the conn, where Volonino and Boyd had also just finished decoding it. Boyd summoned the two young officers forward. Thorson began his recital once more. This EAM was "a valid nuclear control order that authorizes the release of three of *Nebraska*'s missiles." Thorson then read off the target coordinates. The sub was being ordered to begin with a limited nuclear strike.

Thorson didn't believe any of the Tridents would ever fire missiles in anger. If a sub did, it would likely be a limited strike like the one they were now practicing. But Thorson had trouble imagining even that happening. The world would have to have gone mad. If it ever got to the point of detonating these weapons, whatever war they were fighting would have been lost. He had talked about his ultimate mission with Kyung. Thorson had told her he would do his job. But he wanted his new wife to fly to his parents' home in Minnesota, as far away as she could be from where the bombs might fall in retaliation—as if there were any place to escape a nuclear war.

"I concur," Davis said after reading the message.

"I concur, Captain," Boyd said next. The control room was silent except for the four men at the conn who talked in low voices. Volonino and Boyd had top-secret manuals, binders, and code books opened and spread out over the lip of the railing around the conn.

"I concur," Volonino finally repeated. The idea made a great movie plot, but it seemed to him practically impossible for a rogue commander, or even a rogue commander with several conspirators, to launch the weapons without authorization of the President. At least four people on board had to turn keys in different parts of the ship. One of the keys was locked up in a safe and no one on board had its combination.

Could the crew circumvent the security? Of course they could, Volonino knew. His A-gangers had hammers, picks, and blowtorches they could use to break into the safes holding the captain's indicator panel keys. For that matter, he didn't even need the keys to launch the missiles. His men were skilled technicians who knew practically everything about the *Nebraska*'s electronics. They could drill screws out of cabinets, open them up, and hot-wire circuits for the keys like car thieves.

But that would take a lot people being in on the conspiracy—dozens of them in the missile control center and the missile compartment and at the conn. At the

very least, more than a third of the men on board
would have to be involved. And even that might not be
enough. There were other sailors scattered throughout
the sub who, on their own, could flip a switch or pull a
lever to prevent a launch. Volonino would have to
brainwash practically the entire crew into doing something they knew was seriously wrong. He found it im-
practical in the extreme.

"Request permission to authenticate the message,"
Thorson said. Davis and Boyd both concurred, so
Volonino ordered: "Very well, authenticate."

"Authenticate, aye," Thorson responded.

Real SAS code cards were expensive and weren't
wasted for training drills. So for the simulation, they
used a green card with a piece of tape sealed over it,
which Thorson now peeled off to read the series numbers and letters. The alphanumeric code on the green
card matched the code in the EAM.

"The message is authentic," Thorson said.

"I concur," Davis said.

"I concur," Boyd said.

"I concur," Volonino said. "The message is authentic."

If the *Nebraska* actually was ordered to launch nuclear weapons, there would be one more verification
step for the commander. This one was a sanity check
he had to make in his own mind. The Fail Safe Commission had worried about another nightmare scenario.
The encrypted communications net the Pentagon used
to transmit a launch order was secure and the sealed
authentication system seemed foolproof. But no radio
net could be 100 percent safe, and every security system had a weakness. Computer hackers were becoming more ingenious by the day. What if one of them

could break into the Pentagon net and somehow broadcast an emergency action message to a Trident with authentication codes that matched the captain's cookies?

The commission wanted one last safety check. If the world was at peace and an EAM came to a Trident out of the blue, its captain was now under orders to rise to periscope depth and, even though it might put his sub in danger, break communications silence to radio the Strategic Command to find out if it really meant to fire these weapons.

Volonino didn't need any commission to tell him to do that. For the use of these horrific bombs to be even a consideration, his radio shack would be receiving weeks, if not months, of informational EAMs explaining how tensions had risen. There would be military posturing on both sides, there would be heightened Defcons, there would be clashes with conventional armies and many lives lost before someone got the nutty idea to use nuclear weapons, he thought. Maybe in the days when Nikita Khrushchev pounded his shoe on the table, a bolt-out-of-the-blue attack was a realistic scenario. But certainly not now.

If he was just sailing peacefully under the Atlantic and if the only messages coming into his radio shack were the football scores, of course he would have reservations if an order suddenly arrived to launch his missiles. Big reservations. Obviously such an order wouldn't make any sense, particularly if what he could sample above the water told him that life seemed normal.

He was sure his first thought would be "Something has gone horribly wrong here." That's not to say that he would refuse to launch. But Volonino had already

decided that before he did, he would risk an enemy
vessel detecting him so he could radio his bosses to
ask, "Is this what you really want me to do?" A simple
yes wouldn't be enough. He would need some con-
vincing before he launched a missile. His bosses expected him to ask questions. His operating manuals
told him to use his common sense. The Navy didn't
want a Trident captain sitting out in the ocean with his
finger on a hair trigger, thinking that he was expected
to default to a nuclear war.

Volonino ordered Kinman to prepare the sub for the
simulated launch. The lieutenant reached for the overhead microphone. "Man battle stations, missile, for
training without guidance with launcher," Kinman's
voice boomed over the sub's speaker system. "Simulate spinning up all missiles."

1:40 P.M.

Behind a closed door with a sign that warned unautho-
rized personnel to keep out, the missile control center
already had become quietly frenetic. Located off the
port passageway on the sub's second level, just ahead
of the missile compartment, MCC was the nerve center
for the launch. Cape Canaveral crammed into a room
no larger than a den.

A dozen crewmen fanned out inside it, some hooking phone headsets around their necks to test communications circuits, others darting between rows of
refrigerator-size computers with clipboards to take
readings, others hopping into seats before control panels to begin pushing buttons.

Launch order manuals, marked CONFIDENTIAL, were

dumped onto the tops of cabinets and consoles and opened so officers and sailors could read off checklists. They now had less than a half hour to fire the first nuclear salvo.

Fred Freeland finished connecting his headphone and began spinning the dial on a small rectangular combination safe that sat on a ledge almost head-high between the fire control console on his left and the launch control console on his right. With the dial turned to the last number in the combination, Freeland opened the door to the safe and pulled out a ribbed pistol grip painted red with a trigger that had a white plastic guard looped over it. The pistol grip was connected to a thick black spiral electric cord.

Freeland reached inside the safe again and grabbed the tactical mode key, one of the four sets of keys needed to fire the missile. He handed it to Hardee, who for the launch served as his assistant weapons officer and now stood in front of the launch control console. Hardee pretended to insert the green key into a slot at the bottom left corner of the console. If this had been a real launch, he would have twisted the key to the left, to the attack mode.

Reading from a thick manual he cradled in one arm, Hardee pressed his microphone to his lips and keyed it. "All stations, launcher," he said, the order being transmitted throughout the missile compartment. "Make ready reports."

Hardee wanted to be sure everyone was where he should be for the launch. Crewmen began calling him from the command and control center and from different watch stations in the missile compartment. Every man was in place.

Over a loudspeaker in the missile control center

came Volonino's voice: "Set condition one-SQ for training without guidance with launcher. This is the captain. This is an exercise."

At their lowest stage of alert, missiles sat inert in their tubes at a condition level designated as four-SQ. One-SQ brought the rockets to the level of readiness needed for immediate launch.

Freeland repeated the order in the missile control center: "Set condition one-SQ for training without guidance with launcher."

Mark Lyman pushed the "twogle" button on the fire control console to Freeland's left. That setting—training without guidance with launcher, or TWOGL, as it was marked on the button—would simulate the "spinning up" of the missiles for this drill. The fire control console ran the computers that rapidly fed millions of bytes of targeting, fuzing, and start-up instructions into the missiles just before launch. Before blasting off, each missile had to be spun up, which took at least ten minutes. During that time, the gyroscope inside the missile's inertial navigation unit was put in motion and aligned so it could begin sensing the rocket's position in the Atlantic and its movement once it launched.

Lyman was a petty officer first class in the missile division. He had shaved his head for this patrol, and, with his dark mustache, he looked like Oliver Hardy. The thirty-year-old petty officer had spent twelve years in the Navy. Lyman's fire control console consisted of a large tan panel tilted back slightly with four long columns of square buttons—twenty-four in each column—that controlled the twenty-four missiles as they progressed through the launch sequence.

On the panel to Lyman's left was a large video

screen with two paper printers. Sean McCue, one of
the senior missile chiefs, now swiveled a padded chair
around to it. McCue had rushed down from the control
center to take up his duties in the missile center. The
printers were connected to the data recording system,
which electronically monitored the missiles and everything the technicians did with them. Over the course of
a patrol, the recording system spit out about seventy-
five pounds of printouts, which were boxed up when
the sub returned to port and shipped off to a think tank
that evaluated the performance of the weapons and the
crew. Congress ordered an audit of each patrol and
wanted to make sure it was getting the biggest potential bang for the bucks.

As the computers counted down the spin-up, Freeland began receiving instructions from Volonino over
his headphone on the targets for the first three missiles.
"Use the training target package," the commander told
him. For this exercise, all the missiles would be aimed
at a spot in the middle of the Atlantic Ocean. For missiles that would be aimed at countries, Freeland had
top-secret notebooks full of "footprints"—the euphe-
mism the missile men used for targets that could be
destroyed.

"The firing order will be four, twenty-three, and
ten," Freeland announced.

"The firing order will be four, twenty-three, and
ten," the men in the missile control center repeated like
a Greek chorus. The missiles in tubes four, ten, and
twenty-three would be fired.

Even during this grimmest of exercises, there could
be gallows humor. Back at the Kings Bay base, a sign
that hung in the Trident Training Facility's room where
mock missile launches were practiced read: "We de-

liver in 30 minutes or the next one's free—with a mushroom cloud."

Freeland had learned to live with the jokes. He had wide eyes, a Roman nose, and short blond hair. Now barrel-chested from lifting weights, he had been the quintessential nerd in high school. Avidly interested in computers, physics, and math, he had been a member of the chess and science clubs. At the University of Michigan, he double-majored in physics and astronomy and wrote his senior thesis on quasar clustering in the Virgo Constellation.

Freeland's weapons department was large, with fifty-five men. Its missile technicians were some of the most focused on the ship, with little patience for mistakes. They liked to think that they were the only reason the *Nebraska* was out here. The rest of this sub was just a taxi with a nuclear reactor for an engine. It made them seem like prima donnas at times, or so Weller, the chief of the boat, thought.

Freeland was the best-grounded of any of the *Nebraska*'s young officers in nuclear strategic theory. Before joining the sub, he had spent two years at the U.S. Strategic Command in Omaha, assigned to rewrite the Joint Emergency Action Procedures manual. The JEAP spells out how Trident crews should handle emergency action messages, how they should deal with garbles and ambiguities, how they should fight a nuclear war. Its text changes every year. By the time Freeland left STRATCOM, the procedures he had drafted looked almost entirely different from the ones they had handed him his first day on the job. He had also helped formulate some of the new fail-safe measures he now followed as the *Nebraska*'s weapons officer. The STRATCOM work had been fascinating. It had given him a new perspective on how Tridents fit into U.S. nuclear strategy.

It had also been eye-opening to be around the admirals and generals who visited STRATCOM to practice fighting a nuclear war. They would issue their orders for a mock atomic strike, and invariably the next question out of their mouths was: "How much time do we have if we change our mind and don't want the war to happen?" Freeland always answered politely: "If you don't want it to happen, don't start it in the first place." There was no guarantee an attack order could be reversed. The question was understandable but it always unnerved him.

Freeland signed the target package specification sheets he had received from Volonino. "Shift to the training target package," he said to Lyman, handing him the sheets with the launch information for the three missiles.

Lyman leaned back and passed off the sheets to a sailor who worked for him, repeating the order: "Computer operator, shift to the training target package."

The computer operator in this case was Paul Dichiara. With his thick black mustache and dark hair nearly shaved in a short crew cut, Dichiara looked like a Corsican guard. His father was Sicilian-American and his mother hailed from French and German stock. Dichiara had come to the Navy relatively late in life. He had been a store manager for the Blockbuster Video chain until age twenty-five, when his job was phased out. Three years later, he had joined the Navy. Now a petty officer second class, Dichiara, at thirty-one, was older than most of his bosses. So far he had been mature enough to accept it.

Dichiara quickly reviewed the paperwork Lyman

had handed him to confirm that he was receiving the proper targeting instructions. It had Freeland's, Hardee's, and Volonino's signatures on the bottom, which Dichiara quickly verified, plus the date and time the missiles were to be launched. The target data were a bunch of numbers and letters. Dichiara was skilled enough and had enough security clearances that he could have figured out from the codes what country, what city, what population would be wiped out. But he had never wanted to. Let someone else worry about who was going to die.

What Dichiara did worry about was following his small part of the firing procedures to the letter. No new target package was typed in unless it was within the date and time window called for by the EAM. He had found that he had to be anal in this job. The correct target packages had to be entered into the missiles' computers and typed in *correctly*. No mistakes. You don't want a 475-kiloton nuclear warhead hitting the wrong spot.

Dichiara turned around and walked quickly down the first row of large computer boxes to the one that had a blue video screen with a keyboard under it, on which he now typed to program into the fire control computer the targets the missiles would strike. Behind the launch and fire control consoles stood two rows of six-foot-high metal boxes, housing two independent computer systems to talk to the missiles. If one computer system failed, the other was ready as a backup to feed targeting and launch instructions.

Next to Dichiara's video screen was a printer, which now typed out the data he had just entered into the computer. He ripped off the computer paper and walked it over to Freeland at the consoles. "Weapons, automatic printout," the petty officer said crisply.

Freeland took the paper and reviewed its list of footprints to make sure they were the ones Volonino had ordered for the three missiles. They were. The lieutenant had a messenger take the sheet up to the conn so Volonino also could verify that Dichiara had programmed into the computer the targets he wanted struck.

Freeland then turned back to a panel just above the trigger safe. That was where the intent word display was located. Beside it was a thick red metal tube on a small chain that had almost two dozen connector pins at its tip. Freeland took the tube, plugged it into a receptacle on the intent word display, and twisted it to lock it in place. That completed another electronic circuit, which sent a signal to the missile warheads, ordering them to detonate when they reached their targets. If the intent word plug wasn't connected, the missiles would carry duds.

Prepping each missile was only half the job. Getting the tube that it sat in ready for launch was the other half, and just as complicated. While Lyman stayed busy at the fire control console, Kevin Jany was furiously punching buttons and flipping switches at the launch control console to Freeland's right, which monitored the tubes. A petty officer first class just like Lyman, Jany had grown up on an Illinois farm with no idea that one day he would be an electronics technician for strategic missiles. He had already bought forty-one acres of land in Montana to live on when he retired from the Navy. He'd build a house on that land then, maybe keep a couple of horses and a cow, and enjoy the view.

Jany's fingers danced over hundreds of knobs, switches, and tiny square buttons that flashed red or

green lights. Beside him on a small cabinet sat a black
laptop computer, which he consulted throughout the
tube preparations. Keeping twenty-four space rockets
in a pristine state, ready to launch at any time on a half
hour's notice, was a monumental engineering feat. The
buttons, knobs, and lights on the launch control console
were divided into rows of twenty-four. They monitored
or controlled air pressure and temperature inside each
tube, along with the opening and closing of the hatch
and access doors for each one. Side panels on the console contained indicators for outside sea pressure and
temperature, jettison switches in case a missile caught
fire and had to be ejected from the sub, row after row of
alarm lights, plus dozens more switches to control hydraulic valves and the pumping of gases into the tubes.

Volonino phoned Freeland with the depth at which
the sub would hover for the launch. The men in the
missile control center could feel their room tilt up as
the *Nebraska* rose. Jany quickly began punching buttons on his panel to pump nitrogen gas into the three
tubes being used for the launch. He was forcing the nitrogen in to make the pressure inside the tubes equal to
the sea pressure outside. The reason Jany had to equalize the pressure was that when the sub's heavy top
hatch over each missile tube was opened for the
launch, the tube was still covered with a light blue
fiberglass dome to keep the rocket inside dry. But the
fiberglass was relatively flimsy. If Jany didn't equalize
the pressure, the weight of the water would crush it and
damage the missile.

"Weapons, launcher," Hardee told Freeland when he
saw that Jany had finished the pressurizing. "Launcher
ready."

Freeland phoned the missile compartment with the

time window the sub had for getting off its first salvo of nukes.

Back at the conn, one level up, Volonino was now satisfied the first three missiles had the proper targeting instructions entered into them. He had sent Thorson and Davis down to the missile control center with the combination the emergency action message had for the CIP key safe. The key was the last electrical interlock needed to fire the missiles. When Volonino stuck it into the captain's indicator panel and turned it, he was giving his permission for the launch.

As Thorson and Davis hurried down the steps, Volonino grabbed the overhead mike and keyed it. "Entry into the training CIP key safe has been granted," he said over the speaker system. "Disregard all training CIP key safe alarms." The alarm in the missile control center always blared when the real safe or the training safe in the room was opened. Jany and Lyman, who were now designated as the two guards for the safe, recognized Volonino's voice, so they would disregard the alarm and not challenge the officers.

Thorson poked numbers on the keypad lock by the door to the missile control center, then opened it with Davis behind him. Once inside, the two officers walked over to a small white training safe on a perch to the left of the three large safes that actually held the CIP keys. Printed on the little white box was TRAINING CIP KEY SAFE. Looking first at the numbers on the sheet he held, Thorson began spinning its dial left and right.

The training safe's door finally opened, the alarm sounded, and Thorson reached in to grab the training key, which was attached to a green lanyard. But as he pulled the key out he dropped it on the floor.

Thorson bent down to pick it up, but the key slipped

out of his hand again and fell to the floor.

"Jesus, Chad, what's the matter with you?" Hardee said, laughing at his clumsy colleague.

His face flushed, Thorson reached down once more and this time held the lanyard in a tight grip. Davis also grabbed the lanyard. The two officers, now holding the key between them, walked side by side out the missile control center's entrance into the port passageway.

The two men looked a little ridiculous, both of them still holding the CIP key, walking down the passageway and back up to the conn as if they were in a sack race. But everything the crew members did with the missiles they did in pairs, for two reasons—to make doubly sure each task was performed correctly and so each man could watch the other for security purposes. Two-man teams decoded the EAMs. Two men carried the message from the radio shack to the conn. Two men held the CIP key when it was walked from the missile control center to the command and control center one level up. If one carried the key and the other didn't and just walked beside him, there was always the chance the key carrier might drop it into a crevice or slip it into his pocket, and the other person might have his head turned the wrong way and not notice it. The key could be intentionally lost to halt a launch.

The two-person rule was taken to equal extremes during repair work on the weapons. Two trained technicians were always assigned to maintenance chores— one to work on the missile, the other to watch the first man work. The jobs could often become uncomfortable and a bit intimate. To install new guidance equipment in one of the rockets, two men had to squeeze through one of the hatches and into the cramped tube. The installer had to crawl in on his back while the watcher poked his head inside to make sure the first man performed the work correctly. The installer also had to make sure that the watcher didn't sabotage the equipment. If the installer couldn't see the watcher while he worked, the watcher had to keep his hands gripped on the installer's legs—that way the installer could feel if the watcher reached to tamper with something. For some repairs inside the missile tube, one man would even have to lie on top of the other so both could see that the job was done correctly.

Dichiara warned Freeland that they were nearing their launch window. The lieutenant watched the seconds tick off.

One level up, Volonino and Petty Officer Ed Martin stared intently at the digital time readout. The second the launch window opened, Volonino pretended to insert his CIP key into the captain's indicator panel and turn it. He then flipped a training switch on the panel that sent a "training permission to fire" signal to the missile control center.

"Phone talker to weapons, the firing window is open," Volonino told Martin. "You have permission to fire."

Martin phoned the order to Dichiara.

"The firing window is open," Dichiara finally announced.

"Window is open, aye," Freeland acknowledged.

"Permission to fire?" McCue asked.

Freeland verified he had the permission.

"Initiate fire," he ordered.

The missiles now had to run through two more quick phases before Freeland could pull the trigger.

"Denote four, denote twenty-three," McCue announced into his microphone. His order was now carried to other parts of the sub through the speaker system.

Lyman reached up with his right hand to push buttons that put the missiles in tubes four and twenty-three into the denote phase. The long columns of light switches on his fire control console began blinking. During the denote phase, which lasted about forty-five seconds, the fire control computer continued delivering instructions to the missile. More tests were performed, one of them to check that the inertial navigation units were operating properly. The battery inside each rocket was activated so the missile's electronics could operate on its own power.

The missile control center was now silent, except for McCue calling out the phases. Freeland and other sailors crowded around Lyman's panel to stare at the blinking lights for each missile.

"Prepare four, denote ten," McCue said next.

Only one missile could be put into the prepare phase at a time, the final phase before launch. So while McCue put the rocket in tube four into prepare, he ordered the rocket in tube ten to begin in denote. The men wanted to waste no time running the missiles through the launch process, so while one rocket was put into the next phase another was moved to the previous phase. The window to fire could be short, and the crew was always graded on how fast it could get the missiles out of the sub. If everything clicked, the *Nebraska* at this point should be able to fire all twenty-four of its missiles in six minutes.

The prepare phase lasted about fifteen seconds. Tube four's rocket shifted to internal power. The fire control computer sent final flight parameters to the missile. Its warheads received directions on the sequence for when each would separate from the third-stage "bus" that carried them through space and on how each would then fly independently to its target. Fuzing directions also were transmitted for how each warhead would detonate over its target, whether it would strike the ground and blow up or explode in the air. If this had been a real launch, tube four's overhead hatch would now unlock and open.

Freeland held the trigger grip tightly in his right hand and watched the lights on Lyman's fire control console blink from left to right. A small group of lights arranged in two short columns on the right side of the panel were the indicators for the final prepare phase. The bottom light on its second column finally flashed yellow, the signal that the prepare phase for the missile in tube four had been completed. Freeland now had just four seconds to squeeze the trigger.

He squeezed it.

In an actual launch, that squeeze would send signals to the tube to make several things happen simultaneously. A missile's rocket motors aren't actually fired at first. The missiles are popped out of the sub. The trigger sends a signal to a gas generator, which is at the base of the missile tube and powered by a solid propellant, to fire. The intense heat from that gas generator firing instantly flashes water into expanding steam, which forces the missile up and out of the tube. At the same time, plastic explosives laced around the light blue dome covering the top hatch detonate so its fiberglass fragments out into pie sections and doesn't damage the rocket's skin during launch.

As the missile shoots out of the sub's top hatch, it travels up through the water in what amounts to a giant gas bubble, created by the nitrogen that was under pressure inside the tube. The missile actually expels the nitrogen gas as it travels up, which has the effect of preventing any water from leaking through the bubble and touching the rocket. Because of the extreme pressure created instantly by the gas generator, the forty-four-foot-long missile pops up so fast it completely clears the water. In the next instant, when its gyro senses the missile losing momentum and falling back into the ocean, a signal is sent to the first stage of its rocket to ignite. With a fiery plume on its tail, the D-5 missile is off to space.

"Four away," McCue announced, to simulate that the missile had left the tube. Jany had both hands on different switches to close the valve that had been pressurizing tube four.

The launch steps for the two other missiles continued. McCue called out the phases as Freeland squeezed the trigger and the other sailors quietly pushed buttons and followed the flashing lights. With a grease pencil, Lyman jotted down on the glass covering the bottom ledge sticking out from his panel the number of each missile fired.

"Prepare twenty-three."

"Twenty-three away."

"Prepare ten."

"Ten away."

The firing of the first salvo had been completed. A minute later, Volonino's voice came over the loudspeaker.

"Attention, this is the captain. The *Nebraska* has just launched three of her D-5 missiles. This is a limited strike to contain the war and limit escalation. Let us hope that is the case."

4:21 P.M.

It wasn't. In training drills, submarines rarely confine themselves to shooting off just a few missiles. Which is just as well. If a nuclear exchange ever occurs with Russia, there is probably little chance of its remaining limited.

Rich warm smells from the galley, which was preparing the evening meal, began to drift back into the missile compartment. They made stomachs growl. The mock nuclear war had consumed four hours. After the first launch, the *Nebraska* immediately went into evasive maneuvers to escape what, in a real war, would likely be a determined enemy at that point—with the advantage of being able to pinpoint the submarine's position when the missiles fired off. Greg Murphy and Jose Victoria walked down the starboard passageway of the missile compartment's second level. Murphy was one of the roving guards in the compartment. Victoria was one of the senior petty officers in the missile division and its troubleshooter during missile drills.

A bonging bell alarm sounded. Volonino's voice came over the loudspeaker again. The launches would continue. Containment had failed. The *Nebraska* had been ordered to fire a huge salvo of eighteen missiles—or about a hundred nuclear warheads—at the enemy's military and industrial capability. In other words, nuke them to the stone age.

"Set condition one-SQ for training without guidance with launcher," Volonino ordered over the speaker. "This is the captain. This is a training exercise."

Murphy rushed to a large metal power box painted tan, which stood along the starboard passageway, and turned three black knobs. They activated the detonator power supply that would blow the fiberglass covers off the top of the tubes after the hatch opened. Victoria sprinted up to the sub's top level to look for the captain.

He found him at the conn, with Boyd, immersed in a stack of opened manuals and binders. Victoria's other job during a missile drill was to get one of the sets of keys needed to launch—the twenty-four firing unit keys that had to be inserted into the tubes.

Volonino looked up from his paperwork and again handed Victoria twenty-four training keys, which for the exercise were just plastic tabs. The real keys Volonino kept locked in his stateroom safe.

"Twenty-four firing unit keys," the commander said as he gave Victoria the bundle of tabs on lanyards.

"Twenty-four firing unit keys, aye, sir," Victoria repeated and hopped down the stairs.

Almost winded, he ran to the center passageway of the lower level of the missile compartment. Two long rows of giant missile tubes, all painted orange, towered on both sides. At each tube stood a large gray cylinder, which housed the gas generator that flashed the steam to pop the missile out of the sub.

Jose Victoria had been born in Cali, Colombia. His family had emigrated to the United States in 1969, to Queens, New York, where Jose graduated from high school and attended two years of community college. He had had to become a naturalized American citizen to work with these strategic missiles.

Victoria was focused now on becoming a chief in the Navy. He was a petty officer first class and had taken on extra duties aboard the sub—like the first lieutenant job, which meant nursemaiding the young sailors who hauled lines topside when the boat left port or docked— so the command would notice him. It had. Before the patrol, he had received his third Navy Achievement Medal. Work hard, stay busy, keep focused, and you will be rewarded, Victoria firmly believed.

Drive and discipline and devotion to God were important to him. He had been raised Catholic and his South Carolina wife, Deanna, was Southern Baptist, but they both had now become fundamentalist Christians, and committed ones at that. Deanna home-schooled their five children. They felt the public schools had become too secular, too liberal. Not enough attention was paid now to the Bible and prayer and creationism—the traditional beliefs they felt their children would need to make it in life.

Victoria walked briskly down the center aisle of the missile compartment's lower level, stopping at every tube. On top of each tube's gas generator was a firing unit that had a key slot and black ejector ready switch. For a real launch, Victoria would insert the firing unit key into the slot and turn it to arm the gas generator. When Freeland squeezed his trigger, it would send a 2,500-volt charge to ignite the solid rocket propellant in the gas generator and produce 86,000 pounds per square inch of pressure in about a half second. The flash of steam would make a brief but very loud hissing noise, shooting the missile out of the tube almost in an instant. Sailors staring at the rocket through upper-level windows on the tube would see it there one second and gone the next. The force of the 130,000-pound missile being hurled out would cause the sub to dip, so the missile men would feel as if they were in a sky-scraper's express elevator coming to a stop at a bottom floor.

For the training exercise, Victoria just pulled out the small black knob with his fingertips, which made a clicking sound. When he finished at the back end of the two rows of missiles, Victoria grabbed a nearby mike and phoned his supervisor two floors above. "All firing units are armed," he said.

On the second level of the missile compartment, Keith Williams prepared to bounce from tube to tube like a pinball, making sure they were "breathing" properly for the launch. "Breathing" was the slang the sailors used for adjusting the air pressure in the tube so it equaled the sea pressure outside. Computers usually controlled the breathing, but if the computers failed, Williams and other missile technicians stood by, ready to perform the job themselves.

In the second level of the missile compartment, a team of three technicians was on call to quickly fix any problems that cropped up with the twelve missile tubes in the forward section, while another three-man team kept watch over the twelve tubes in the aft section. As each missile went through its spin-up, denote, and prepare phases, the three men in each section danced among the tubes in a careful choreography ready to manually override mechanical, pneumatic, and electrical functions. The computers almost always worked properly, however, so most of Williams's time during launch exercises was spent just staring as the missiles automatically went through their phases.

Williams had been in his rack reading one of Tom Clancy's *Op Center* thrillers when the announcement came to spin up the missiles. He was in a sour mood. The reason he was reading the Clancy book was that a chief had rousted him out of the enlisted lounge, where he had been watching a movie, *The Faculty*. No movies during missile drills, the chief had ordered, and especially not while the squadron inspectors were aboard.

"Hell, one of the inspectors was in there watching it with me," Williams groused. It was only the first week and already he was tiring of this patrol.

Yeah, there were a lot of things Williams would change with this Navy. And he wasn't shy about letting people know what they were, which at times hadn't won him popularity contests on the sub. He had been an Air Force brat, which probably explained why he had no accent. Moving constantly had wiped out any sign of his roots. An African-American, he had a deep and loud baritone voice that didn't seem to go with his slight frame. Williams had joined the Navy more as a fluke than out of any desire to follow his dad's footsteps. He had made good grades, but school had bored him. He enjoyed chemistry and electricity, but it had only gotten him jolted from playing with electric sockets and practically blown up from home chemistry labs. Now he was stuck with these missiles for at least another two years.

The way they treated nubs on this sub rankled him the most. He had arrived as a petty officer third class and had to put up with what seemed to him like juvenile fraternity hazing—the worst racks, the dreariest watches, hassling when you tried to grab an extra dessert in the chow line, sailors above you making it hard to get your qualification card signed for your dolphins.

Of course, it didn't help that he was brash and opinionated. And it didn't matter to him if the person listening was a sailor or a senior officer. Williams gave them all an earful. Lieutenant Commander Ashton, the old executive officer, had invited him to his stateroom when he first came aboard and had asked him what he thought of the qualification ritual for receiving dolphins.

Williams told him he thought it was "all bogus."

"Once you're qualified, you'll change your opinion," Ashton answered, taken aback.

Williams didn't, and he told Ashton that after he had earned his dolphins. In fact, as a protest, he had intentionally slowed down having his qualification card signed so he would pin the insignia on during his second patrol, not his first. That made a few people on the sub angry. The old salts lectured him that earning his dolphins was an important tradition, to make a sailor knowledgeable about other parts of the ship or able to save another sailor's life in an emergency.

Williams thought it was all bunk. A nub could put out a fire just as well as someone wearing dolphins. The qualification only made you *familiar* with the rest of the boat. It didn't mean you could actually run other parts. If he and all the other missile techs suddenly died, those nukes in the back wouldn't have a clue how to fire these rockets.

The Mickey Mouse stuff stopped once Williams had those dolphins. People who had harassed him suddenly became his friends. Then on his third patrol he was promoted to petty officer second class. Wow! What a difference. He went from being a know-nothing sailor to someone who commanded respect. Williams wasn't impressed. His pay changed, but he hadn't. Before, his complaints had been passed off as whining. Now they became "valid points." But they were the same damn complaints. Go figure.

Williams hadn't decided whether to reenlist. He might try to become an officer. Then he could make some changes around here. . . . Well, maybe not right away. Ensigns were listened to about as much as new sailors. But somebody had to buck the system or things never got changed. Whatever, he'd be a different kind of officer, he promised himself. That old saying that the chiefs ran the Navy was a bunch of crap, he thought. The "blue shirts," the enlisted men under the chiefs, they were the ones who ran these ships. He'd treat those young sailors with more respect.

Williams was now part of the aft team of technicians watching the missiles. Each tube was surrounded by a labyrinth of pipes, valves, levers, gears, pumps, wires, cables, plugs, and an assortment of electronics boxes whose job was to keep conditions perfect for the missile to launch at any time. He stopped at one of the tube's electronics boxes, a gray metal indicator panel, so he could verify that the maintenance switch was set to normal, that the failure indication lights weren't blinking red, that the missile gas valves were lined up properly. Then he checked to make sure all the green lights on the panel were glowing green.

The tubes he oversaw were all breathing properly on their own, so he didn't have to pressurize them himself. Williams wasn't all gaga around these missiles. He'd been living with them so long he'd practically forgotten what they contained inside. They became just a collection of giant metal cans, as far as he was concerned. His friends back home thought it was cool, his working with strategic weapons. He found it dull. Maybe these monsters had been important twenty years ago. But with the Cold War over? Come on . . . it was overkill.

The missiles began their prelaunch phases.

"Denote thirteen, denote two," McCue's voice from

the missile control center came over the loudspeaker. Whooshing sounds came from the tubes as they vented air.

"Prepare thirteen, denote fourteen."

Williams stood at tube thirteen. In a real launch, he would be ready with a tool to mechanically unlock and open the tube's top hatch if the computer failed. All he had to do now was watch.

"Thirteen away. Prepare two. Denote three."

The rocket in tube thirteen had just simulated being launched. Williams now had to gag the missile gas valve. He quickly pushed a switch on the tube up and slid a lever to the left. For a real launch, he would verify that valves operated properly so seawater that rushed into the empty tube didn't back up into open pipes.

The valves worked. Williams shouted "Gag!" and hurried to the next missile.

McCue continued the sequencing of the launch phases.

". . . Five away. Prepare seventeen. Denote nineteen."

With headphones clamped to his ears and a microphone under his chin, Stacy Hines paced back and forth on the starboard passageway of the missile compartment's second level, between the CAMP and a large white bulletin board just to the right of it. The CAMP was the control and monitoring panel—four banks of switches, lighted buttons, and glowing red digital readouts Hines used to make sure tubes pressurized properly and valves opened and shut when they were supposed to. The white status board had columns of boxes for each tube, which Hines marked as the missiles marched through their phases.

Hines supervised all the technicians racing about the compartment. He was like a football coach roaming the sidelines, talking on his phone to the missile control center in the front of the sub and barking out instructions to his technicians running between the tubes.

McCue and Hines were the missile division's two top chiefs, with polar opposite personalities. McCue was intense and quick-witted. Hines was a good ol' southern boy, firm with his men, but still laid-back. Outside the Navy, the loves of his life were his family, his white cockatoo, Casper, and building furniture. In that order. When his wife, Tammy, now wrote him, she made sure to include lines about how their daughter, Amanda, missed him, an update on Tammy's pregnancy with her second child, and—mandatory in every letter—whether Casper was okay.

With McCue spitting out numbers rapidly over the loudspeaker, it was easy to get mixed up on what phase each missile was at in the launch sequence. Hines had his own way of tracking the tubes on his status board. He used the scoring for a dart game he had learned in Scotland, a game called cricket. When a missile went into denote, he marked a slash in its box. For prepare, he crossed the slash to make it an X. When the missile was launched, he circled the X.

The launch sequence neared its end. By his count on the status board, Hines had three more missiles left in this salvo.

"Twenty-one away. Prepare twenty-two."

"Twenty-two away. Prepare eleven."

"Eleven away."

"Set condition four-SQ," McCue announced on the speaker system. The three missiles not fired in the simulated launch would be spun down.

"Safe all fire units," Hines called out to his technicians.

Victoria turned the three knobs on the detonator power supply box off, then sprinted down to the missile compartment's lower level to push back in the black button on each gas generator's firing unit.

SUNDAY, 5 P.M.

The command and control center on the sub's upper level had been eerily quiet through much of the missile drill. Ed Martin had stood over the captain's indicator panel console, using a grease pencil to line out the numbers on the panel for the missiles that had been fired. The CIP console had two screens that flashed different-colored lights as the missiles spun up, then went through their denote, prepare, and away phases. The left screen contained twelve rows of square lights for the twelve odd-numbered missiles. The right screen had the rows for the even-numbered missiles. The keyhole for the CIP key was located at the bottom of the console and was now covered by a Plexiglas seal held in place by a yellow plastic tie that would have to be cut to slide the seal off. For an actual launch, Volonino would turn the key left, from the HOLD position to FIRE.

During missile drills, Martin left his reactor operator job and monitored the CIP console. Watching the square lights blink across each row, he was another set of eyes making sure the missiles moved through their prelaunch phases smoothly. Even in training drills, every time the final AWAY light blinked red, Martin felt a tingle inside. There was no "OOPS!" button on this panel. No way to call these rockets back.

Martin knew full well the overwhelming power of

the weapons. In addition to his preacher's license, he also had a bachelor-of-science degree in nuclear engineering. On long boring watches back in the engine room, he had plenty of time to think about the *Nebraska*'s ultimate mission. So did the other nukes. The conners couldn't be bothered with it and the officers avoided nuclear war conversations during wardroom meals, but the nukes were more introspective, he thought. When they ran out of other subjects to discuss, they would have philosophical debates about the morality of killing so many people. Sometimes they would paint pictures in their minds of the aftermath. They also tried to picture what would happen if the sub failed in its ultimate mission. Martin didn't think any of the nukes would freak out if they received an actual order to launch. But he was sure there would be dead silence, save for a lot of them murmuring to themselves: "Oh my God, I don't believe this is happening."

Thorson and Davis finally approached the conn with the tenth emergency action message. This one was the order to terminate the launch. The message was valid. Their pretend war was over.

"Captain, we're satisfied," Hunnicutt told Volonino. The squadron inspectors had seen enough. "You can secure the exercise."

"Aye, aye," Volonino answered and began closing his manuals. "That, as we say in nuclear warfare, is that."

15 • Angles and Dangles

Mark Nowalk was worn out. It was Monday night and he could count on one hand the number of hours he had slept in the past two days. He expected to stay busy as a nub, but at this pace, he worried he would collapse before the end of the first week of patrol. The ensign looked around the control center, as if he were plotting an escape. His watch over, a dinner in his stomach, no one in sight who might bug him with another chore—he savored the thought now of six uninterrupted hours to himself.

"I'm going to my rack," Nowalk said, to no one in particular.

Kinman, who was now the officer of the deck, heard him and looked up from his paperwork at the conn.

"What, to stand on your head?" Kinman answered with a laugh.

It wouldn't be far from the truth. The sub was about to play angles and dangles. Nowalk gave Kinman a defeated look. Sleep was out of the question for a while, unless he did want to stand on his head. Angles and dangles would make sure of that.

In the thick of combat, a submarine has to be able to dive quickly and deeply into the ocean or shoot up to a shallower depth, particularly if it is evading an enemy torpedo. Volonino wanted to limber up the *Nebraska* by porpoising it in a series of dives and ascents. Tilting up and down also would test how well the crew had stowed away gear. If a cabinet was going to tip over or a wrench was going to slip off a shelf and clatter on the deck, Volonino would rather cringe over the noise now than during a quiet game of rabbit and wolf with another sub.

A messenger passed by the conn and asked Kinman if he wanted a drink from the galley.

"Yes, a cup of coffee," the lieutenant deadpanned. "Shaken not stirred." Alcohol is not allowed on U.S. Navy ships.

Kinman reached up and grabbed the microphone. "Rig ship for deep submergence," he announced over the loudspeaker.

Throughout the sub, sailors had been pushing drawers into cabinets, strapping down boxes, shutting open doors, putting tools away, and cleaning off tables. The angles would be steep and equipment could go flying if not anchored down.

Volonino walked briskly into the control center and plopped in his seat at the conn. Kinman gulped down the last of the coffee the messenger had brought him.

"Chief of the watch, report if the ship is rigged for deep submergence," he ordered as his skipper looked on.

One of the machinery sections plus the engineering department hadn't yet phoned that they were ready, the watch chief answered.

Minutes later, the machinery and engineering spaces called. They were ready. Angles and dangles began.

Standing up near the edge of the conn, Volonino ordered the planesmen to send the sub to 650 feet, at a twenty-degree angle. Quentin Albea and Jason Bush, who sat before the two steering wheels, shoved them forward.

In less than a second, the *Nebraska*'s nose began to dip. As the control room tilted down, Volonino, Kinman, and the other men standing up leaned back. Newcomers to the sub always reached for posts or overhead bars to keep from tumbling over. The submariners had practiced the drill so many times, they knew how to tilt back slowly with their arms folded so they could remain standing as the boat pointed down.

Shawn Brown, now the diving officer, called out the depth change every fifty feet until the sub finally settled at 650. The men straightened up in unison. Beasley, whose job now was to be Brown's messenger, wandered in with a cup of hot cocoa he hoped to sip between errands.

"This is not the time for confectioneries in here," Volonino snapped. "Take it out and dump it."

The chiefs glared at Beasley. The last thing they needed was cocoa flying. Mystified, he carried his steaming cup out of the control center.

Kinman ordered Brown to take the sub back up to 225 feet, again at a twenty-degree angle. The men in control leaned forward, like snow skiers jumping off the end of a ramp.

The sub leveled out at the higher depth. The men straightened up again. Kinman ordered another plunge.

"Full dive, both planes," Brown commanded his two men at the wheels.

The *Nebraska* headed back down to 650 feet, but this time at a steeper angle, twenty-five degrees. The men leaned back, the rubber on their running shoes gripping the deck enough so that they didn't slip and fall on their rears. They looked like slapstick characters in a Mack Sennett silent movie.

At 500 feet, Brown ordered Bush and Albea to pull their wheels back, hoping the lead time would be enough so the boat would settle at 650 feet. It wasn't. Brown overshot the depth by thirteen feet.

Up the sub went again to 225 feet. In the radio shack just forward of the control center, Horner broke into a limbo dance. Angles and dangles could be fun for the sailors, although sometimes the clowning turned dangerous. On some subs, the more daring sailors in the missile compartment sit on folded wool blankets at the back end of a center passageway, then bobsled down to the front when the sub dives—a risky run because they might slam into sharp metal corners.

"I want to do some three-dimensional stuff," Volonino decided. He ordered Kinman to sink the boat to 750 feet at an even sharper angle, thirty degrees this time, and bank it to the right as well—a maneuver the sub might take to evade a torpedo.

The control room tilted way down, like a dump truck emptying its load. If you grasped an overhead bar, it felt like hanging from a tree limb. On the ride back up this slow-motion roller coaster, the clanking sound of metal striking glass could be heard from one floor below. A silver pitcher from the battleship *Nebraska*'s tea service had tipped over in its case in the wardroom. Volonino frowned. At least it was happening now, when they were just practicing. The mess stewards would have to Velcro it down more firmly.

By 8 P.M., buckets of popcorn had been made, pitchers of soda had been poured, and crewmen settled down in different parts of the sub to "burn a flick," as they called it. The officers crowded into the wardroom to watch a sex-and-action film, titled *Wild Things*. The sailors in the crew's mess showed considerably more taste and slipped *Pleasantville* into the VCR. In the chiefs' quarters, Weller sat shaking his head. The sub had six hundred videos on board—some of them so new the distributor hadn't yet put them in stores—and these guys were watching a dull chick flick: *You've Got Mail*.

"I can't believe it," he muttered. "You all act like you got channel fever."

A bad movie was called a "hacker." Every now and then a sailor would intentionally pop a hacker like *The Little Mermaid* into the VCR, just to irritate everybody. But the rule was if you burned a hacker you had to stay and watch the entire film.

The patrol was beginning to quiet down. The shake-down period had ended. The frenetic pace of the previous week eased. Hardee now stood at the conn as the officer of the deck. He had the midwatch, which lasted until midnight, a time when it was the most peaceful.

The *Nebraska* was sailing back to Kings Bay and would dock in port for thirty-six hours. Volonino wanted to repair the towed sonar array that had snagged. But the more important piece of equipment that needed to be fixed was in the "deeper," the data processing equipment room. The room had two main-frame computers (one a backup), which ran the naviga-tion planes and the sonar as well as monitored the sound level of all the machines on board. The comput-ers were the brains of the sub's electronic network. But the power supply for one of them was broken, making it unusable. The sub could stay at sea with just one computer, but if it also broke, the *Nebraska* would be "the Helen Keller of the sea," as the technicians put it, with no sonar to detect other vessels. It would have to surface and sail home.

The crew was delighted to be heading back to the base and—for those not stuck in the sub on watch—to be able to spend at least a couple of nights with families or girlfriends. Weller was always uneasy about returning to port at the beginning of the patrol. "Just another opportunity for sailors to get into trouble," he said.

Hardee glanced at the digital display indicator on the first row of consoles in front of the conn, looking for any bright white lines down the green sonar waterfalls that might indicate a vessel was near them. There were none, and had been none for several hours. That was usually how the midwatch became—slow, with nothing going on.

Reggie Rose stood behind a young seaman at the inboard planesman's seat who was learning how to steer the boat. Rose whispered instructions and occasionally leaned over to turn the wheel for the apprentice when he drifted off course. Much of the night had been spent with trainees at different watches so they could gain experience in the control center. The instruction always made Volonino nervous. It was like teaching teenagers to drive the family car.

A sailor with a clipboard and pen walked up the port steps to the control room just to the left of the conn. He stopped at the top of the stairs, reached for a green knob on a metal box to his right and turned it counterclockwise, then made a note on his clipboard. The knob was the reset for the "beast buoy." Its military designation was the BST-1. A sailor was assigned to stop by the box every half hour to twist its green knob or check that the knob had been twisted. If the knob was left unturned for a total of ninety minutes, a buoy broadcasting an emergency radio signal was automatically shot from an opening on the side of the sub.

When it sailed on strategic alert or during combat, the *Nebraska* would not transmit radio messages to headquarters for fear that the enemy would pinpoint its position. If the sub suffered a catastrophic accident or if it was destroyed in combat, the beast buoy would be the only way the Navy would learn that it had lost one of its Tridents. As a torpedo was about to strike the sub in battle, for example, the officer of the deck was under orders to rush to the beast buoy control and, in his last act before dying, to flip a switch on the box that would launch the buoy. The Navy had had its scary moments when a sub accidentally launched its buoy. Admirals scrambled, terrified that one of their Tridents was sitting in pieces on the ocean bottom, until the boat finally radioed back to headquarters that it was still alive.

Hardee leaned an elbow on the railing around the conn and stared at nothing in particular, lost in his thoughts. For the next three months the midwatch would be like this. Rarely would the sub rise even to periscope depth. "Nothing special," he said to himself. "Just poking holes in the water." It did give the men a chance to get to know one another. But it wouldn't be long before boredom set in. By the end of the patrol, they would all be white as sheets of paper, angry with each other, and ready to get off this pig.

PART II

. . .

THE NEXT TWO MONTHS

16 • Free at Last

After a little less than two days in port—enough time to fix both the towed sonar array and the power unit in the data processing room—the *Nebraska* again pushed away from the pier at the Kings Bay Naval Base on Thursday morning, May 13, and set sail for the Atlantic. There would be hardly any port calls for the next seventy-eight days, and that had the crew grumbling. Since the end of the Cold War, the Tridents had begun pulling into more ports overseas and along the American coast. This gave wives or girlfriends a chance to fly to wherever the subs tied up for a short visit with their loved ones. But this patrol would be mostly an uninterrupted and lonely one for the men.

The drive out the Cumberland and St. Marys channels was uneventful this time. Thorson, for the first time, served as the officer of the deck, standing on the bridge as the submarine steamed to the edge of the Atlantic's continental shelf, where it was deep enough to dive. He was thrilled. This steel behemoth now turned right or left at every command he issued to the control center. A surge of pride coursed through him when he thought about how far he had come from that tiny town in Minnesota. What he was experiencing now would seem completely foreign to his family and friends back home.

With the ocean bottom now hundreds of feet below them, the crew dismantled and hauled down the gear from the bridge once again and slammed shut the top hatches. Volonino allowed Thorson to remain as the officer of the deck for the dive underwater, which again was his first time at that job as a fully qualified watchstander.

This was an even more thrilling experience. Yet through it all, the pain of just leaving Kyung one more time kept gnawing at him. What was she feeling now? Would she be lonely? Would she be safe? He couldn't stop worrying, even with the many other things he was supposed to monitor as the sub sank smoothly into the ocean.

The planesmen pulled their steering wheels back and the *Nebraska* finally settled at a depth of 160 feet. Thorson let out a long breath. There had been no problems with the dive, as far as he could see. He hoped nobody had sensed how inexperienced he felt on his rookie run as officer of the deck.

Al Brady, the sub's navigator, had been frantically busy, bent over the channel charts in the control center, during the sail out. And he had no time to catch his breath now that the submarine was submerged almost fifty miles off the Atlantic Coast. No time at all. Brady rushed back to the navigation center, the large room behind the conn. He had to scramble for the next twelve hours. Lieutenant Brady was looking at seventy-eight days of steaming in the vast Atlantic Ocean and he hadn't put a pencil to a single chart to plot where the submarine would sail.

Brady began spreading maps of the ocean out onto the large plotting table in the nav center. He had a reason for being late in marking down the sub's course. A fundamental tenet of the Navy's security procedures for its ballistic missile submarines is that nobody onshore—absolutely nobody—knows where each vessel is hidden in the ocean. That makes each boat impervious to espionage. The exact course a Trident sails is left entirely up to its captain. The top-secret Patrol Order Volonino had received before the *Nebraska* drove away told him the areas of the ocean the Strategic Command wanted his submarine to be in during different times of the patrol. But those areas were huge—as much as forty thousand square miles—so if a spy intercepted the Patrol Order, his Navy would gain very little that would help in finding the sub.

Before the *Nebraska* set sail, Brady had walked into Volonino's Kings Bay office and closed the door behind him, and the two men had talked generally about where the skipper wanted the sub to roam in the operating areas. Volonino could travel in them as he pleased. They discussed when the sub would be on strategic alert ready to fire its missiles at a moment's notice, when the alert level would be lower and the crew could make more noise, when Volonino would want to travel fast and when he would want to slow down. It was all big picture. Nothing of the conversation was written down on paper. Brady left the session with enough of an idea from Volonino that he could have begun marking up his charts. But Navy regulations barred him from drawing a single line until the sub was under way.

Brady and three assistants now hurriedly pushed rulers and protractors back and forth over the charts, marking and measuring. They had to work fast. Volonino didn't allow the sub to sail in the ocean without a course he had approved. To get them going, Brady had to quickly plot the first chart, which represented only a tiny piece of the overall operating plan.

At the same time he had to draw the plots for future days. All totaled, he would probably mark up as many as forty charts during the patrol.

Deciding where to sail wasn't easy. The Navy's maritime charts are suited more for surface ships than subs, which have to navigate three-dimensionally under the water. Brady had to plot a path for the sub, a 3-D shoe box, which was wide and high enough so the officer of the deck would have room to maneuver in it safely if he had to.

Each Trident captain picks a random pattern to sail in his operating area. But none wanders aimlessly. Brady had dozens of considerations in coming up with a route. The sub couldn't stray out of range of the communications transmitters posted around the world. Sometimes Volonino would want his sub sailing so covertly that even his own Navy wouldn't be able to detect the vessel. Underwater topography was crucial. Brady didn't want to hit any mountains. At the same time, he could use the ocean's ranges, slopes, underwater currents, and warm or cold eddies to acoustically mask the sub. Naval Intelligence had told him where unfriendly subs might lurk, so he could avoid them. But sometimes commercial shipping lanes are a good place to hide, because you can get lost in a noisy crowd.

Twelve hours later, an exhausted Brady had finished marking up the first chart for the route that day as well as the overall patrol plan for the entire seventy-eight days. Volonino reviewed the charts, tweaked the plan for different days, and finally approved it. Fewer than ten percent of the men aboard the submarine were cleared to see the navigation charts and to know the *Nebraska*'s exact position, and even fewer knew the entire operational plan.

Volonino kept a "cartoon chart," as he called it, posted on a bulletin board in the hallway outside of the sonar shack. It was a map of the entire Atlantic Ocean, with the sub's position plotted on it—except that that position was about a hundred miles off from where the Trident really was. The cartoon chart gave the crewmen a general idea of where they were in the Atlantic. Even if one of them revealed the coordinates after they left the sub, it wouldn't be spy-quality data.

The accurate chart was spread out on the plot table just to the right of the conn. It was never covered, but if a crewman not authorized to see it wandered over for a peek, the navigation technicians, always hovering there, would shoo him away.

Every night, Brady found a secluded spot on the sub to boot up his laptop computer and type out a top-secret summary of what the *Nebraska* had done during the day, where it had sailed, and the operations it had conducted. Volonino reviewed and edited the Patrol Report every week. It became one of the most sensitive documents composed aboard the sub, a detailed diary, in effect, that was delivered to the Strategic Command when the Trident returned from patrol. Even paper trash left after the final draft was printed was considered highly classified and destroyed. For good reason. The Patrol Report each Trident delivered to STRATCOM recounted in detail where the sub had been.

But even that report would be of only limited value to a spy. Though every other Trident captain considered the same factors Volonino did in planning his patrol and they all followed the same rules for staying stealthy, no two captains ever took the same route. The Navy had conducted post-patrol surveys and found there was enough variation in the path each commander plotted that an enemy force would never see a pattern.

Volonino felt free at last. He had finally "broken the gravitational pull," as he liked to say. It was a frustrating process every time, a struggle he always had with man and machine to get the deployment under way. An avalanche of problems could weld the ship to the pier if he wasn't determined to escape: medical problems with his men, mechanical problems with his equipment, supply problems getting spare parts.

The Navy refuses to let a Trident push away unless it is 100 percent ready, which meant Volonino would never get to sea if he weren't fanatically determined to overcome the thousands of niggling problems. Once he was out in the ocean, it was a different story. If a piece of equipment broke, he had millions of dollars' worth of spare parts and expert repairmen aboard who in most cases could fix it. If not, he had backups. He could continue sailing, even if everything wasn't running perfectly. It was his call. He ran the show. Finally, he was in complete control of his destiny.

Volonino had a few days on his own before headquarters wanted him for missions, so he began training his crew relentlessly. Kinman staged nonstop drills. Fire drills, flooding drills, simulated casualties, faulted dives . . . they went on and on. Then Volonino began "mini-wars"—mock torpedo battles and simulated missile launches using the Indy computer program to generate the enemy. For one exercise, Kinman used the war in Kosovo to concoct a scenario. NATO bombers had accidentally destroyed a vodka factory, which had enraged the Russians, he announced over the sub's speaker system. The Tridents had been put on a heightened Defcon. (To keep the exercise politically correct, Kinman used fictional names that rhymed with Russia and Kosovo. Life had been simpler when you knew your enemy, he mused to himself.)

Instead of steaming east toward the middle of the Atlantic, the *Nebraska* was ordered to sail south near the Florida coast, to play the role of an enemy sub for another Trident from Kings Bay that was taking its tactical readiness exam. Before playing rabbit and wolf, however, Volonino stopped off at Port Canaveral and picked up twenty-nine Naval Academy midshipmen out for a day cruise. The Tridents considered "midshipmen operations" only slightly less important than their strategic mission. The submarine force was short of officers. The underwater Navy competed fiercely with its aviation and surface branches for recruits. Volonino had ordered his crew to turn on the charm.

The sub puttered off the coast for a day, then loaded the students onto a tug that took them back to the coast. Volonino decided to keep the boat surfaced overnight. Just before dawn the next morning on May 27, sailors climbed up the ladder, carrying apple turnovers and mugs of hot coffee, and crowded onto the top deck to watch a space shuttle launch. Volonino had positioned the sub so the shuttle's rockets arced and separated over its stern. The view was breathtaking.

But Volonino had to hustle his men back inside the sub and submerge it quickly to make his next appointment: rabbit and wolf services for the other Trident. He always tried to show professional courtesy when he served as the mock enemy for another boat taking its tactical readiness exam. He was under orders to be aggressive during the exercise. At the same time, he usually knew the skipper of the other sub and didn't want

to embarrass him when he was being graded. The
worm could turn, Volonino knew, and that skipper
could one day make him look foolish on his exam.

Of course, Volonino couldn't help it if the other sub
dumped a victory into his lap. Or, to be more accurate,
dumped it into the laps of a very excited Kinman and
Habermas. Volonino normally had more senior men
serve as the officer of the deck and junior officer of the
deck during rabbit and wolf exercises, but this time he
allowed the two young lieutenants to take the conn.

"Okay, this is stupid," Kinman finally said to Habermas, rolling his eyes. They had spent hours hunting to
no avail. "We're never going to find this guy."

Habermas stared at the digital display indicator in
front of the conn, looking for any sign of sound on the
green waterfalls. The rabbit and wolf game was being
played in a fifty-by-fifty-mile box. When a Trident
wanted to really hide, it could disappear. No way two
rookies like him and Habermas could find an experienced skipper in this large an area. The two lieutenants
made their obligatory runs through the box anyway.

But suddenly a faint line appeared on one of the waterfalls. The chances of its happening were one in a
thousand. The other Trident came rolling out of the
dark shadows. By dumb luck, Habermas had pointed
the *Nebraska* in a direction so that it had practically
bumped into the other Trident without her knowing it.

The two lieutenants excitedly summoned Volonino
and Boyd to the conn to show them their prize. The
commander and his executive officer sat dumbfounded.

"I can't believe it," Volonino said, chuckling.

For the next forty-five minutes, Habermas and Kinman kept the *Nebraska* behind the other Trident, in her

baffles, where she couldn't hear at first that she was being followed.

Finally, Volonino ordered Habermas to send the Trident a coded signal that she had been shot. It wasn't his style to give a fellow skipper this kind of bloody nose, particularly when that skipper had squadron brass aboard grading his every move. But this poor soul had no hope of escaping.

Doc Philbin wasn't having as much fun back in the pharmacy. At first he thought this patrol would be different from the others. So far, the two weeks at sea had been peaceful—stitches sewn on a couple banged heads, Motrin for an occasional cold, powder for athlete's foot, nothing serious. Then it came. By the end of the third week, it came like gangbusters. He shouldn't have expected that this patrol would be different.

When the hatches were closed for the last time, the sub became a temporary incubator for any bacteria or viruses the crewmen had brought aboard. After several weeks, the air filtration system eventually cleansed the atmosphere inside of unfriendly germs. But until then, no one could escape another person's bug, and the men invariably came down with head colds or stomach ailments in droves. They called it the "patrol crud," and during the first two weeks of most cruises, Philbin dispensed as many as five hundred cold tablets.

The bug was late coming this time, but when it did, it swept through the sub with a vengeance. By the end of the third week, Philbin started seeing the first symptoms—nausea and vomiting. He administered shots of Phenergan, often used for airsickness, to stop men from throwing up. Philbin gave so many injections the crewmen started calling it "shot-in-the-ass disease." But as the stomach virus spread—becoming more virulent, or so it seemed to its victims—the bug got a new nick-name: the "one-eighties." Not only were men puking their guts out, they now had diarrhea as well. The torture lasted for twenty-four hours. Kinman, one of its victims, had never felt so miserable being flushed out from both ends. The only time he climbed out of his rack was to rush to the john for one or the other of the one-eighty.

At least forty men came down with the dreaded dis-ease. There were probably more, but Doc became so overwhelmed sticking needles in asses he lost count after forty. Finally, on the sixth day of the plague, Volonino decided to fight back.

"We're sick of being sick," he announced defiantly on the sub's speaker system. He ordered every man not hugging a commode to drop what he was doing and grab a sponge.

"I want everything on this boat wiped down," Volonino commanded. So far, he had avoided the one-eighties. Enough tours on subs had made him religious about tak-ing his vitamins, washing his hands constantly, and avoid-ing sneezing men. But at this rate, there'd be nobody left to drive the ship if he didn't kill this little critter.

Spray bottles full of antibacterial antiseptic were passed out, and for three hours men wiped down door-knobs, galley equipment, shower stalls, toilets, sinks, anything humans touched. The wipers started at the bow of the boat and worked their way to the back. Everybody cleaned. Nobody complained. They were all on a crusade to rid their sub of this ugly enemy.

It worked. Philbin had a couple more cases of the one-eighties. But by the eighth day, the bug had been killed. The crew was saved—or at least until the next newcomer came aboard carrying a new disease.

17 • On Guard

Hardee was becoming bored playing with the airmen. It was a scientifically proven fact that submariners were smarter and better-looking than aviators—or so Hardee and every man in the underwater service would swear on a stack of Bibles. And the moron who was flying the P-3 Orion over the *Nebraska* certainly was living up to the submariners' opinion of aviators.

The P-3 is a propeller-driven aircraft always based along a coast. It has radar and acoustic sensors, plus twelve men, crammed inside it. Since the end of the Cold War, the P-3s have spent many hours chasing Iranian subs and drug traffickers speeding in cigarette boats. But the plane still practices its original mission of hunting and killing sophisticated subs, like the ones the Russians sail. Which was what the P-3 flying over the Trident now was practicing. Or, to be more accurate, trying to practice.

June had just arrived and the *Nebraska* was in the middle of its fourth week on patrol when headquarters ordered it to play antisubmarine warfare games with the P-3. Bigwigs from the squadron had come aboard the week before to test how prepared the crew was to take its tactical readiness exam, which came at the end of the patrol. The squadron officers had come aboard more for CYA, the crew suspected. They administered a far more grueling battle test than what the crew would ever see on its tactical readiness exam from the Atlantic Fleet officers. That way, if the crew flunked the fleet exam later, the higher-ups in the squadron had the paperwork to show the Navy that the crewmen had done fine when they checked them.

The *Nebraska* had also picked up more midshipmen, who were tugged out to the sub when it had surfaced briefly off the coast—nine of them this time. They would spend almost a month aboard as part of orientation cruises the academy had its students take in the summer. Behind their backs, sailors called them "ifnags" for "ignorant fucking Naval Academy grads," but Volonino ordered each person assigned to a petty officer first class who was supposed to qualify them on junior watches and sell them on the submarine service. This batch seemed eager to learn the boat, and Volonino thought he could persuade several to join.

Today's antisubmarine warfare game was becoming an exercise in futility, as far as the P-3 was concerned. The surveillance planes are usually pretty adept at hunting subs. They can sneak up on a Trident in the wink of an eye using their fancy electronic gizmos and nail the sub if it isn't careful. But this P-3 was having a bad day. It might as well have been a Delta Airlines jet flying up there trying to find the *Nebraska*, for all the luck it was having. The sub had risen to periscope depth and stuck its periscope and practically every other mast out of water—the only thing left to do would have been to hoist an American flag—and still the hapless plane couldn't find it.

Hardee had begun needling the pilots over his radio. The P-3's call sign was Red Claw. Hardee began addressing them as Bear Claw, the name of a pastry sold in the 7-Eleven stores back home.

"No, no, Red Claw, Red Claw," the pilots kept correcting him. But Hardee kept misstating their call sign.

Finally, the P-3 aviators, who were becoming increasingly irritated with this smart-ass in the phantom sub, gave up.

"Request you give us your latitude and longitude," one of them radioed Hardee so the P-3 could rendezvous over the Trident.

Hardee darted over to the charts on the plotting tables to his left in the control center. He quickly jotted down a list of numbers, hopped back up to the conn, grabbed an overhead mike, and read them to the aviators.

Silence on the other end of the line.

The coordinates Hardee had given them were for an airstrip at Mayport, Florida. It was the naval air station where the P-3 was based. Maybe these guys could at least find their way back home.

Hardee guessed that the pause in the transmission was the time it took for the P-3's navigator to plot the coordinates and then for the pilots to cool down.

"Request *your* latitude and longitude," the voice of one of the pilots came over the radio speaker in the control center, each word enunciated slowly with a "you-asshole" edge to it.

Volonino, who had been sitting in his stateroom and listening in on the control center conversation from his speaker, bounded up the stairs to the next level. He was furious. "Bear Claw" he could tolerate. But rubbing their noses in it with coordinates to their own base was too much. Volonino chewed out Hardee, then radioed the P-3 himself to apologize for the first set of numbers. But everybody else in the room thought it was hilarious, including Hardee.

It was the evening shift on June 5, and Matt Suzor was trying to keep his mind on the navigation chart spread out over the plotting table in the control center. But it was hard. The petty officer second class was in charge of marking the charts with every move the sub made as well as plotting its future course for the night. But a minute didn't go by that he didn't wonder what was happening hundreds of miles to the west of him.

Finally, Volonino walked into the control center, from its forward entrance. He had just made a stop at the radio shack. He strolled over to Suzor's plot table and, smiling, placed the message under his nose. It read:

> Paula Suzor delivered a 6 lb. 15 oz. baby girl at 12:47 p.m. on 5 June 99 at Camden Medical Center, St. Marys, Ga. Daughter's vital statistics are: name: Sarah Elise Suzor; length: 19 1/2 in.; hair color: blonde; color of eyes: blue. Mother and daughter are doing well. Paula states they love you, are doing well, she looks just like you and Sarah was in a hurry to get here.

Suzor folded the message and slipped it into his pocket. He was overwhelmed with joy and excitement. His third child had been born. Three girls. Sarah had arrived a week early. (Quickly, he would later learn. Paula had been in labor for only ten minutes.) But elation quickly turned to depression. He was not there for the birth. And Suzor wouldn't see Sarah for at least another month.

*   *   *

Several days later, the *Nebraska* finally went on strategic alert. It had been loitering in the Jacksonville Operating Area off the Florida coast when its Patrol Order went into effect. Gradually the Trident slunk off, like a cat slipping out the back door into a black night. Its level of operational security, the stealthiness with which it traveled underwater, began to increase. Before it had been sailing at Opsec Delta, which was the most lenient and which allowed the sub to transmit and surface. As the Patrol Order took effect, so did Opsec Charlie. The sub could no longer transmit and no longer surface. The only vessels that could detect it were other U.S. Navy warships.

Then, as it crept farther away from the American coast, the *Nebraska* moved into Opsec Bravo. Now it had to remain undetected even by its own Navy. (Later in the patrol it would practice operating at Opsec Alpha, an even more covert posture to be used during wartime.)

By the end of the first week on strategic alert, the *Nebraska* was thousands of miles from home. The waters where it roamed were as deep as eighteen thousand feet. At long last, it had begun the mission that ballistic missile submarines have assumed for four decades: nuclear deterrence, standing quiet sentinel under the ocean. Kinman had developed a reverence for this part of the patrol. They were playing a big boys' game. They were off to find their hiding place where nobody else on earth could see them.

The *Nebraska* would remain on alert for almost a month, creeping along quietly at about three knots in its giant operating area, hidden in blackness from the rest of the world, just listening on its radios but not transmitting. Ten of the eighteen Tridents in the fleet roam the Atlantic and Pacific at any moment. Different combinations of submarines are at different stages of alert. Some are on what is called "mod-alert" and can make more noise so they can train at other tactical jobs, such as hunting other subs, as the *Nebraska* had done during the beginning of its patrol. At mod-alert, Volonino could be ready to fire his missiles within three hours if he had to.

Meanwhile, other Tridents would be at full strategic alert, sailing silently and ready to fire quickly. However, none of the Tridents on strategic alert now remain at that level for as long as they did during the Cold War. When the Soviet Union was the enemy, each ballistic missile sub spent many more weeks on strategic alert. But even then, it was overkill. All ten Tridents wouldn't likely have been needed to fire all their missiles at once at the Soviet Union. The Pentagon's war plans envisioned starting a nuclear conflict with a limited strike, not a massive retaliation. And today, a massive retaliation is even less likely. Any conflict now would almost certainly escalate slowly, the admirals believe, so there would be plenty of time to bring other subs up to higher alert levels.

Kinman still ran training drills for the crew during the time on strategic alert, but he had to simulate more actions—such as pretending to run the diesel engine instead of actually cranking it up—to avoid making noise. The *Nebraska*'s routine became less frenetic, more predictable, and, for many in the crew, more boring. Kinman had found that the atmosphere on board became stiffer in some ways during strategic alert. Superiors and subordinates became more formal with each other. This was their real job. This was why they were out here. This was the mission. They weren't just training or marking time. They were now on guard. If they failed, if they lost communications with STRAT-COM, if they accidentally broached the surface, if they were detected by another vessel, there would be serious repercussions back home. Kinman always found it to be a lonely month.

It was one of the loneliest times for Volonino, as well. There were no more games of rabbit and wolf with other subs, no more radioing messages to planes in the sky, no more calls to his superiors for advice or guidance. The Trident was now operating independently of any other vessel, U.S. Navy or otherwise. Volonino had broken the gravitational pull, but he was also on his own. Any problem that came up during the next four weeks he would handle himself. There was no outside help. And he was the man the crew now looked to for every answer.

The good news was that he didn't have to have a perfect solution for every problem. He just had to avoid making bad decisions. This was the ultimate fishbowl, except that the fish were on the outside. He was on trial the entire time. Every man in the sub was watching and measuring him every day. He'd be a fool to believe that *everybody* aboard liked him. He was sure that a few of the men had it in for him. One of them could accuse him of misconduct, and, true or not, it could derail his being promoted to captain. But he had twenty years of experience as a submariner, which he was pretty sure was enough to keep him out of trouble. He didn't have to worry about being a "wise" commanding officer or a "beloved" one, whatever that was. He just had to avoid being a stupid one.

There were, however, rewarding aspects to being in this fishbowl. This was also *his* world now. That was one of the things he liked about submarine command. Never again in his life would he have this kind of control over his environment or over a group of men. He would dearly miss that when he had to give up command of the sub.

At this moment, however, Volonino was worried far less about nuclear deterrence than about what Marvin Abercrombie was pissing into the toilet.

It had started several days before, the pain in Abercrombie's lower back. He had thought it had just been caused by the stress of the endless training drills the squadron riders had put the sub through. He was glad that was over. Abercrombie had a lighter schedule this patrol. He had been the chief of the A-gangers until Bob Lewis had come aboard about two months ago to take over that job. After this patrol, Abercrombie was transferring to another assignment, but for now he was aboard to help Lewis with the transition.

The squadron riders' leaving brought a present for Abercrombie. The tugboat that had come out to pick them up also carried with it mail for the submarine, and in that bag was a letter from his wife, Wanda, informing him that he was once again a grandfather. Hanah Renee had been born on May 31. His third grandchild. At forty-five, Marvin Abercrombie was one of the elders of the submarine. He had salt-and-pepper hair, a mustache, and a stocky frame. He had played the sub's Santa during Christmas cruises.

He was also a romantic. On the overhead section of his bunk was a collage of pictures of his wife and his children and their spouses, plus his grandchildren. Marvin and Wanda had known each other in their Tampa, Florida, high school, but they hadn't started dating until he had graduated and joined the Navy. Wanda had long red hair and light brown eyes that seemed to reach out and grab him. Two weeks after they first went out, he asked her to marry him. She said yes. Marvin was nineteen and Wanda was just seventeen. But they were in love, and over the next twenty-six years—through all the moves, all the skimpy paychecks, all the separations while Marvin was at sea—that love had only grown deeper. Every chance he got on patrol, Marvin wrote long passionate love letters to his wife. During the previous patrol, to celebrate their twenty-fifth wedding anniversary, Marvin and Wanda had renewed their vows on the top deck of the *Nebraska* just after the sub pulled into port. Ed Martin, the boat's licensed minister, wore his black vestment and performed the second wedding.

Abercrombie wished he had Wanda by his side now to nurse him. The pain in his lower back had grown worse over about five days, then it had spread to his groin. He had tried exercises and miles on the running machine, but that didn't help. Even more worrisome, his urine had turned pink, like fruit punch. It was blood, Abercrombie knew. This was serious.

By ten o'clock at night on June 11, he could stand it no longer. Abercrombie hobbled back to the pharmacy on the second level of the missile compartment to find Doc.

Philbin had just finished giving what he hoped would be the last shot of Phenergan in a bottom for a case of the one-eighties when Abercrombie walked in. His skin was pallid and sweating and he was in intense pain.

"What's going on?" Philbin asked calmly, pushing the granny glasses up on his nose.

Abercrombie described his symptoms.

"On a scale of one to ten, what's the pain like?" Philbin asked.

"Nine," Abercrombie answered.

Philbin suspected immediately that the chief was passing a kidney stone, particularly after Abercrombie told him he had been peeing blood for almost a week.

Philbin took a urine sample. It now looked like red-tinged coffee. The corpsman hooked Abercrombie up to an intravenous bag so he would receive fluids and gave him painkillers, then he rushed to the front of the sub, where he found Volonino at the conn.

"Most likely it's a kidney stone," Philbin told the skipper. "The symptoms are classic.

"But I'm not convinced it's just a stone," Doc warned. The symptoms could also be for more serious ailments, such as nephritis or a form of cancer. Volonino would have to break radio silence and ask the submarine force's Atlantic Fleet headquarters if it wanted Abercrombie evacuated.

Gone are the days when no medical or personal crisis would interrupt a ballistic missile submarine's strategic patrol. There are still enemies out there, but the admirals recognize that the atmosphere isn't dire enough that a sailor with a life-threatening illness has to remain aboard. Ailing crewmen are routinely medevaced, and if the sub is near a port or tugboat, sailors are often allowed to leave in order to deal with family emergencies at home.

The next morning, the *Nebraska* poked a radio mast out of the water and fired a message to the Atlantic Fleet's headquarters in Norfolk, Virginia.

Eight hours later, the video screen in the radio shack

lit up with Norfolk's reply. The Atlantic Fleet's doctors
had reviewed all the vitals and symptoms Philbin had
included in the message. The *Nebraska* was at a spot in
the ocean where it would be extremely difficult for a
helicopter to reach for an air evacuation. Besides, heli-
coptering a patient off a rocking sub on the high seas is
a risky undertaking. There was nothing more doctors
on shore could do for the chief than what Philbin was
already doing on the boat. So keep him there. The doctors did not "recommend medevac at this time," the
message read. It continued with more than a dozen
things the physicians wanted Philbin to do. They included keeping Abercrombie on an IV and painkillers,
sending him to his rack, watching for any spikes in his
vital signs, and having him urinate through a sieve to
catch the kidney stone when it passed.

For six days, Abercrombie lay in his bed in the
chiefs' quarters, the tube from the IV bag stuck into his
arm. Painkillers relieved the backache, but the color remained in his urine. When Sunday rolled around, he
couldn't climb out of bed to go to church services, as
he always did at sea, so the church came to him. The
door to the chiefs' quarters opened and the men from
the wardroom service filed in, filling the room. They
held hands in a circle and prayed for Abercrombie.

"God is good, isn't He," he wrote to Wanda in a letter that night, which he hoped he could somehow get
off the sub and to her.

Philbin took Abercrombie off the IV after a week,
but ordered him to continue drinking plenty of fluids to
flush out the stone. Eventually, the back pain went
away without the pills. Abercrombie might have
passed the stone, but he couldn't be sure. The sieve
never picked up anything. That worried Doc. The

chief's urine still had a tinge of pink. Something else was wrong, but Philbin didn't know what.

Volonino had another worry in the back of his mind. He had limited medical supplies aboard the sub. Abercrombie had consumed many IV bags and Doc didn't have a large inventory left. Volonino constantly played "what if" games in his head during these patrols. What if another sailor became seriously ill? Would he have enough medicines aboard to treat him? Would he have to race to a port for a medevac or deal with the problem once again on the sub? He hoped he wouldn't have to face such questions.

18 • Halfway Night

Show time!

Liebrich wasn't processing messages in the radio shack now. Now he was the master of ceremonies, sporting a beard, wearing a black T-shirt and jeans, and speaking into a microphone. The crew's mess was jammed with rowdy and laughing sailors, chiefs and officers, some of them in civilian clothes, all of them with stomachs full from a rich meal of lobster tails, steak, corn, hot rolls, and decorated cakes. The galley had cooked up the feast. Stacy Hines, one of the missile chiefs, had wolfed down ten lobster tails, the sub record if anyone was keeping count.

The *Nebraska* continued to sail silently underwater, still on strategic alert. But it was time to have fun. To break the ice, Liebrich punched a button on the VCR, and on the wide television screen above it flashed a video movie that some of the sailors had shot.

Volonino, who sat in the front row, cringed. Abercrombie cringed. The movie producers were his A-gangers, not known for G-rated videos. Abercrombie knew the skipper didn't want anything obscene.

The sailors hooted and hollered. The video opened in the "snake pit," the lower-level machinery section where the pipes wound in and out. An A-ganger with his back turned to the camera seemed to be masturbating.

Abercrombie grimaced. "Oh, man, the skipper is not going to like this one," he murmured to himself.

The A-ganger turned around with a goofy smile on his face. He'd been wiping the grease off a grease gun.

Next scene: Steve Dille, the senior chief from the engineering department, began explaining how nuclear power moved the boat. The camera panned to show Dille cracking a whip in the missile compartment while the nukes pulled broom handles like galley slaves on a barge.

Roars of laughter in the crew's mess! Even Volonino began chuckling. Hell, why not. The video stayed PG, and besides, his men were peaking. The workload had been crushing the past six weeks. He'd challenged them to do things they'd never imagined they were capable of doing. The constant combat drills had seemed so oppressive in the beginning, but they now snapped them out to show off their skills. One day, the crew even staged a Damage Control Olympics, where teams from different divisions competed with one another to see who was the best at fighting simulated casualties. Volonino began each patrol worried there might be an accident because the proficiency of the men was low. Now he always worried about a mishap because they were too cocky. But that was a good worry to have. This was always his most satisfying time of command. He felt like a proud papa with all his sons gathered around him.

And tonight was a night to have fun. It was Saturday, June 26, halfway night. Whenever a patrol is half over, every Trident skipper celebrates it by letting the crew unwind. Thirty-nine days down in the *Nebraska*'s

patrol, thirty-nine days to go. In years past, halfway
nights had been raunchy. There were porn movies and
beauty contests with sailors dressed as drag queens.
Many of the submariners missed that. The PC ayatol-
lahs ran the service now. The submarine admirals
didn't want a scandal in their force like the Tailhook
outrage that had crippled the naval aviation community.

Throughout the patrol, the crewmen tried valiantly to
keep one day from blurring into the other. They organized contests and games and special events to break the
monotony and lift spirits. There were cribbage matches
and card tournaments and a casino night. Pools were
held for who guessed the right time that the boat would
dock at port. Everything conceivable was raffled: a
DVD player, camcorder, shotgun, gift certificates to
electronics and sporting goods stores. Hundreds of dollars circulated through the sub during the patrol as the
men bought raffle tickets or won prizes, with a percentage taken off to keep the recreation fund and wives'
club activities afloat. Once the squadron riders left and
there were no other outsiders on board, the men also
could buy no-shaving chits for $10 that allowed them to
grow beards—Kinman tried a goatee this time, which
came out a bright red—and sailors were allowed to
wear different-colored T-shirts under their poopie suits.
On Father's Day the week before, the chiefs had cooked
a prime rib dinner and the men opened cards their wives
had sent before the sub sailed. Another night was designated pizza and ugly tie night; the officers cooked pizzas for the crew with extra toppings and the seamen had
a contest for who wore the ugliest tie. (Jenny Brady,
Al's wife, had bought a bundle of ties at the Salvation
Army for the men.)

But the men looked forward to halfway night as if it were Christmas. Work had been called off that day. The lounges had movie marathons showing one video after another. An auction was held to pick the night's scullery queen, the man—usually the most unpopular one—who would wash dishes during the festivities. The proceeds from the bids went to the recreation committee.

After the homemade video, Liebrich called up the other events. They had a Jell-O-eating contest, an ugly beard contest, and a contest to see who could drink the most water. They played Pie in the Eye (whipped-cream pies were sold that you could smash into anyone's face) and Corn on the Cob (creamed corn that you bought to smush into the chief of the boat's face). Each man opened a halfway night box, filled with candy and cards, his family had prepared before the patrol; Weller had stored the boxes away on the sub. A makeshift rock band played. Abercrombie's church group sang a song. Skits were acted out to poke fun at one another. The nine midshipmen aboard staged a spoof imitating all the officers aboard. One of them strutted around in a poopie suit with glasses pretending to be the captain. Another stuffed pillows in his suit and rolled up his sleeves to play Kinman.

Then everyone settled down for what they all wanted to see the most. Liebrich popped another video into the VCR and the crew's mess quieted. About thirty of the wives and girlfriends of the *Nebraska*'s men had made this movie. They had taped it at Wanda Abercrombie's house. Each woman stood before the camera and told her husband or boyfriend how much she loved him, how much she missed him. Some prattled on, some felt too awkward to utter more than a couple of sentences.

It didn't matter to the crewmen that the women had taped the video more than a month ago when the sub was in port and had delivered it to Weller on the promise that it wouldn't be shown until halfway night. They watched it and smiled at one another and laughed and poked each other in the ribs as if the tape had been magically beamed into the sub that day.

Even when the festivities had been racy, Abercrombie had found them boring. And he wasn't a big fan of lobster tails. The wives' video was the only reason he had come tonight—to see his children and grandchildren and Wanda, who was never at a loss for words. She told him how much she loved and missed him. "I can't wait for you to come home so I can give you a long list of honey-do chores," Wanda said cheerily. Abercrombie smiled. The garage was waiting for him to clean it up.

Thorson had sat near the back of the audience waiting anxiously for this part of the festivities as well. He blushed when Kyung appeared on the screen and playfully reminded him she was still his wife. (The other officers would kid him about that later.) "Don't worry, I'm okay," Kyung finally told him. "I love you and I miss you."

He knew she wasn't okay. She would be all smiles when he returned. But she had cried when he left, and he was sure there had been times during this patrol when she felt bitter that he had abandoned her. Oh, how he missed his wife and having a normal life. He was torn between enjoying his job and having to be at sea to do it. What kind of career was this if he spent so much of it waiting for time to pass? Sure, some days

flew by, but others dragged on forever. And few were
worth reliving or even remembering.

After each wife and girlfriend had appeared individ-
ually, they all came together in a group to say goodbye
and wave. Then the TV screen went blank.

The next night, another feast of roast beef, baked
chicken, pasta, rice pilaf, vegetables, and cheesecake
was spread out, but this one was in a private dining
room of the Osprey Cove. It was one of the nicer restaurants near the Kings Bay Naval Base, just off a golf
course in the quaint little fishing town of St. Marys.

The women now took their turn at celebrating.
About thirty wives and several girlfriends of the *Nebraska* crew gathered in the Cove's dining room to
mark the halfway point of the patrol. They laughed
too and hugged and chatted and gossiped. Two long
tables were arranged for them to sit at for dinner. A
cash bar stood near the buffet table, but many sipped
iced tea, sodas, or water. They all wore their Sunday
dresses. That was one of their traditions. Look your
best, even if he isn't here. The other rule: no kids. This
was girls' night out. Hire a baby-sitter so you can unwind.

Patrols are as lonely for the women as they are for
the men. Perhaps even lonelier. Sailors aboard the
Navy's surface ships were away for just as long or
longer, but at least they can write and receive letters
regularly from home. Warships pull into ports often
during their cruises or bring on replenishments at sea
from supply vessels, so bundles of mail can be exchanged. Seamen on modern aircraft carriers can even
e-mail home or phone loved ones on credit cards.

Not so on a Trident submarine. For almost three

months, husbands practically disappear from the face
of the earth. Port calls are infrequent, mail drops only
occasional. Wives feel abandoned.

There is at least one form of personal communication between couples, but it is one-way and so strictly
rationed it often adds to the frustration of the separation. Each sailor is allowed to receive eight "family
grams" during the patrol. His wife is given a blue form
with forty blocks printed on it, in which she can write
just forty words that will be transmitted by radio to the
sub at intervals.

The women find family grams intimidating. It is almost impossible to say what you want in just forty
words. On top of that, the message isn't private. Three
staff officers at Kings Bay review each family gram before it is radioed to the sub to make sure it is cheery.
The command wants nothing to dampen spirits on a
Trident. Bad news, such as a serious illness or death in
the family, is censored. So are code words, riddles, and
risqué terms.

The wives often seethe over the constraints. Who do
these admirals think they are that they can open your
mail and order you to write only lovey-dovey words?
Why should the truth about life back home be kept
from grown men? Often as not, the truncated messages
raise questions in the minds of the sailors who read
them rather than assure them that everything is fine at
home. It irritates the wives as well that the family
grams are sent off to the ether, as far as they are concerned. Forty words, and you never receive a reply.
Nevertheless, the men look forward to those precious
forty words, and the *Nebraska*'s men complained when
the family grams didn't arrive for two weeks of the
submarine's alert period.

Volonino tried his best to ease the isolation back home. The *Nebraska*'s other crew, the Gold Crew, which was back at Kings Bay training for its next voyage, took care of the Blue Crew's wives while the men were away. Gold Crew sailors made themselves available to drop by a Blue crewman's house to unclog a drain, fix a lawn mower, jump a dead battery, repair a washing machine. It was a reciprocal arrangement. In port, Weller, as the chief of the boat, spent much of his day fielding phone calls from distraught wives of the Gold Crew.

Before the *Nebraska* set sail, Volonino asked every wife and girlfriend to attend a predeployment briefing. He fed them dinner as an incentive to come. At the meeting, each woman received a thirteen-page document with hints on coping during the three months.

This was not a normal business trip their men were taking, the document began. Practically every aspect of the Trident's operation was classified. The women were sternly warned not to discuss with strangers when the sub set sail, when it would return, where it would operate, or anything else they might have overheard their husbands saying about its capabilities or the assignments it might be undertaking while at sea. Secrecy was paramount. The wives all talked among themselves. That was unavoidable. But keep it in the family.

The wives' club that had been formed for the Blue Crew's spouses had a phone tree that quickly spread important information, such as a port call the sub might take. The document had dozens of phone numbers of people on and off base who could help with emergencies, including a checklist with hundreds of items the crewman and his spouse should take care of ahead of time: numbers for all bank accounts, the location of medical records and insurance policies, powers of attorney established, updated identification cards for traveling on base, repair schedules for cars, doors and windows that had to be locked, phone numbers for the plumber and electrician, candles in case of no power, and hurricane evacuation routes.

Volonino ended it with advice to both spouses on readjusting when the husband returned after three months at sea. He listed "things wives should remember":

He may seem different. Think how much you changed. Remember that he's been subject to daily regimentation and routine. He may rebel against schedules and preplanned events. Leave room for some spontaneity.

He may want to celebrate his return with a spending spree. If you can't afford it, hold tight to the purse strings. The urge to spend should pass.

Expect him to be surprised or hurt that you've coped so well alone. You can reassure him that he's loved and needed and that just because you did manage without him, it wasn't that you *wanted* to.

Volonino also noted "things husbands should remember":

Don't disturb a family setup that has been working well without you. Ease into the system slowly. Enjoy being "an honored guest" for a while.

Take it easy on the kids, especially where discipline is concerned. It's best for the kids to have a *consistent routine*, so let Mom's rules stand. Don't barge in as the "heavy."

Don't alter the financial affairs. Chances are your wife's been handling them fine. Remember that prices have possibly soared while you were gone.

Expect your wife to be a little envious of your travels, so go easy on the descriptions of seven-course banquets and beerfests. *Bring her a gift.*

Remember that while you've been active and busy, she has faced all of the home, car, and family's perils by herself. She has earned a break.

Expect her to be different. The fact that she can cope without you does not mean that she wants to. Coping with separation has changed her into a more competent and independent person. Expect this to be permanent and enjoy the fact that you do not have to worry while you are gone.

The dining room began to fill as the women lined up with plates before the buffet table. For Molly McCue, this truly was a night to relax. Her young son and baby were with a sitter. And at least for tonight, she didn't have to cope with problems that cropped up with the women in this room.

Volonino had what other Navy ship captains had: an ombudsman who was the wife of a crew member and who represented the wives before the brass. Molly, who was Chief Sean McCue's wife, was one of the *Nebraska*'s ombudsmen. Volonino had appointed two of them for the sub, and he had chosen them carefully. Volonino treated his ombudsmen as senior members of his staff, and when they spoke, he listened.

A competent ombudsman could save him countless headaches. She had to be willing to be awakened in the middle of the night to aid others in a crisis. She had to be savvy with the Navy bureaucracy and help a wife cut through red tape. She had to keep her mouth shut and not gossip about other people's problems. She also had to be firm with headquarters but not so pushy that he had a lot of angry staff officers on his hands when he returned home.

So far, the patrol had been bearable for Molly. On past cruises, she could count on phone calls from frantic women waking her in the middle of the night. They phoned constantly with financial or household or children problems or just because they felt lonely. This patrol, however, had had few family tragedies. She had even come home several days when there were no messages on her machine from wives.

Molly was well qualified to handle personal emergencies. She had a bachelor's degree in psychology and worked as a mental health case manager. But most of the problems she had untangled so far were nothing more serious than having an error in a sailor's paycheck straightened out or helping a spouse replace a lost military ID card. There had been no deaths in the family, except for two pet dogs. Only two wives had had to go in for surgery, and only one of the surgeries had ended up being serious, a hysterectomy that had developed complications. Molly had arranged for a Red Cross message to be sent to the *Nebraska*. The husband had been sent home.

But Molly was ready for this patrol to end. Was she ever ready. She was worn out dealing with other people's problems, worn out coping with troubles of her own. Caring for a baby and a busy six-year-old boy was a big enough job. There had been ups and downs, but Ryan hadn't acted up as much this patrol as ones past when he cried often and seemed angry at everyone because he missed his dad. She had kept him busy this time, sent him to his grandparents to take his mind off Sean's not being home. But just this past week, one of Molly's best friends had been killed in a car wreck. Her husband also was in the Navy, on a surface ship, away at sea when the accident happened. If only Sean could be here to help her deal with this. Molly wanted him home.

The wives finished their cheesecake and ordered coffee. A television set with a VCR was wheeled in. The husbands had prepared their own videotape before they left port. Now the women sat quietly, some teary-eyed, as one crewman after another appeared on the screen to say how much he missed his wife.

Just as the sniffling became noticeable, however, Chris Wilhoite, one of the sonar petty officers, appeared on the screen in his blue poopie suit. His wife, Melissa, who was president of the crew's wives' club, perked up.

Wilhoite held up a set of skimpy lingerie. "When I come back, I would like you to wear this the first night," he told Melissa. Eyes began to widen in the dining room and frowns turned to smiles.

"Okay, wives, turn your heads," Wilhoite said with a sly grin. "This is for my wife only." Slowly he began pulling down the zipper on the front of his poopie suit. "I know you other wives out there are probably jealous."

Gasps!

Wilhoite let the poopie suit drop to his ankles. On the front of his white boxer shorts was printed HI HONEY.

Wilhoite turned around. On his bottom spread across both cheeks was I LOVE YOU.

Melissa turned red and covered her mouth. The room broke into laughter. Wives doubled over cackling and wiping tears.

The room stayed in an uproar the next five minutes. It was as if a dam had burst, with all the pent-up emotions and frustrations now washed away in laughter.

Wanda Abercrombie always loved the halfway nights for the wives. When her husband, Marvin, appeared earlier on the television screen, it had seemed as if he were looking only at her.

Wanda had long learned to cope with the loneliness of the patrol. This was her nineteenth, and the memories of each had long blended into the others. Patrols like this one that came in the summer, however, had been easier to wait out, she had found. Night didn't come early and she wasn't chained to the routine of seeing her youngest son off to school each morning. They could escape the house and visit relatives in other cities. When it was quiet at night with the children or grandchildren asleep, and her only company was the television set, that was when she missed Marvin the most.

But Wanda was devoted to her husband and the Navy he served. Over the years, she had set her alarm clock in the morning to the number of the boat on which Marvin sailed at the time. It was her little I-love-you to remember him when he was gone. For the past three years, her alarm had been set for 7:39 A.M., the *Nebraska*'s number. She had also learned to become self-reliant when he was away. During the last patrol, she had repaired a garbage disposal by herself.

The Navy wants one big happy family, and Wanda had become one of its happiest members. She had a son in the Air Force and a son-in-law in the Marines. When Marvin's boat pulled into port, she always stood at the dock with a sunny smile to greet him. As long as Marvin enjoyed his work, she would follow him to the next

duty station. Navy wives become as close-knit as their
husbands, for better or worse. Moving every two or
three years, they are always strangers in their neighbor-
hoods, outsiders who come and go like nomads. No one
in the civilian world could comprehend what they go
through—even if they could, the wives aren't allowed
to say much to civilians about what their husbands
do—so the women form their own community within
the community. Some, like Wanda, make the best
friends of their lives in the Navy community. When
Marvin retired, she would miss the Navy as much as he.

Wanda had become a mentor to many of the wives
new to the sub, wrapping an arm around them, inviting
them to her home, introducing them to the strange new
life they would lead. Many of the rookies this time she
had met at an ice cream social the month before to welcome them to the *Nebraska*. She had started a walking
program with some of the women so they would be
slim when their husbands returned. Wanda had served
as an ombudsman on a number of boats and had
chaired countless social committees. She had counseled many wives the past twenty-six years. Once she
had even talked one out of committing suicide. "You
can make it through the patrols," she told them all. Yes,
there were rough spots, periods of depression, but they
went away.

It was almost ten o'clock at night when the dinner
ended. Two-piece boxes of Godiva chocolates had
been passed out to each wife, along with a card her
husband had written before setting sail. Kyung Thorson decided to slip away with some of the wives of the
junior officers for an after-dinner gathering at one of
their homes. Because they were coequals in terms of
the misery they shared, wives were not supposed to

wear their husband's rank when they socialized with one another, but they often did. In fact, they seemed cliquey to Kyung. The enlisted wives had sat together at one table during the dinner and the officers' wives had sat at another section. Even the wives of black crewmen had kept to themselves. Kyung thought they all acted as if they were in high school.

Chad had appeared on the video. He hadn't said much more than the other crewmen. He loved her. He missed her. He hoped she was doing fine in her job. Be patient. The patrol was half over and soon he would be home. To tell the truth, she hadn't become emotional when he came on the screen. It was nice to see his face. But he was gone. She couldn't touch him or talk to him. He seemed more like a memory to her at the moment.

And that made her miserable when she thought about it. She had a love-hate relationship with Chad by the time halfway night rolled around. Some days she loved him. Some days she hated him for leaving. There were so many headaches to deal with on top of work: paying all the bills, making the house payments, trying to regrow the grass Chad had killed on the lawn. She found herself resenting him because he wasn't there to deal with these chores. She was sick and tired of his being away.

It wasn't that she needed to cling to him, certainly not like many of the other Navy wives did to their husbands. Kyung knew she was different from them, and from Chad. When he had to leave, she wanted him to make the break cleanly, get it over with, not say the goodbyes over and over again as he wanted to. She was sure it was her Asian background that made her this way. She wasn't used to showing emotion as Chad did.

His family was all "I love you" and hugs and kisses. Not her. You show love by example, she believed. Actions mean more than words. Chad was somebody she loved. That was why she had married him. But he was not somebody Kyung needed.

Until she went to college, she had been shy and submissive, as Korean girls were expected to be. She had been sheltered by strict parents most of her young life. Her mother had doted on her father, and the family expected her to be the same way with Chad. But university life had opened Kyung's eyes. She had relished being on her own and discovering that she had a right to be treated as an independent person. She had been attracted to Chad because he was so kind and intelligent. She could never have fallen in love with a man she didn't respect, or who didn't respect her.

Now it seemed to Kyung that she was passing between two worlds—the university world she had just left, which now felt alien to her, and the Navy world, which she wasn't sure she wanted to be a part of. It had been hard being with her college girlfriends during Chad's first patrol in 1998, when they had been engaged and she was finishing her pharmacy degree at the University of North Carolina at Chapel Hill. All her friends could talk about was doing this or that with their husbands or fiancés, going to this place or that. She couldn't. Chad was gone. Her friends had no idea what it was like having the person you loved erased temporarily from your life.

But coming to Kings Bay was like landing on the moon. Few of the wives were true career women, as she wanted to be. Few even had lives of their own. The majority lived for their husbands. Here it was the end of the twentieth century, when men and women were supposed to be equal partners in marriages, but Navy wives were still expected to be good Navy wives and support their men at sea. Her job always followed his. She must never say or do anything that hurt his career. Kyung was shocked. All these women talked about was what they could do to help their husbands get promoted, how they would fix up their home at the next duty station.

Other young wives struggling to keep separate careers found Navy life a straitjacket as well. As the patrol dragged on, wives got on other wives' nerves. Many resented that their friends were picked for them. Their husband's circle was supposed to be their circle. Nobody said it out loud, but making friends with outsiders was frowned upon. You could quickly become an outcast if you weren't careful. Just as on the sub, everyone knew everyone else's business at home, and made a point of knowing it. No wonder that the divorce rate was high in the submarine service. The *Nebraska* usually had one separation every patrol. A few husbands strayed during port calls. And a few wives cheated while their husbands were away.

Kyung refused to be part of the service "scene." She had found a job as a pharmacist at a local drugstore, and that kept her busy. She would socialize with the other wives, play the Navy's game as best she could. But she refused to learn all that military jargon—she couldn't believe how some of the wives were so proud that they knew every term—and she wasn't going to talk about the *Nebraska* day in and day out. She was a twenty-five-year-old professional woman. Chad had already been jolted once or twice when she refused to drop what she was doing at work and rush to the pier to be with him. But as she saw it, he was the one who had joined the Navy, not she.

19 • Emergency

It was nearing the end of June and the *Nebraska* had only a few more days on strategic alert. Soon it would sneak back to the Georgia coast and surface to pick up torpedoes—their warheads removed—and practice shooting them at a range off the southern Florida coast.

The world above the Trident had just provided another reminder for why the ballistic missile submarines still remained on strategic patrol. The week before, American F-15 jet fighters had intercepted two Russian strategic bombers flying near Iceland. The two TU-95 Bear warplanes, part of a massive military exercise staged by the Kremlin, had flown to within striking distance of the United States. The bomber incident came several weeks after two hundred Russian troops had raced into Kosovo's capital of Pristina to seize its airport against the wishes of NATO peacekeepers.

Word of the Bear incursion first came into Volonino's radio shack via one of the weekly news reports the sub received. It was soon followed by intelligence reports and analyses with more details on exactly what the bombers were up to.

The Russian gambit made Volonino pause. So far there had been three "strategic events" (his term) on this patrol: the standoff with Russia over Kosovo, the bombing of the Chinese embassy, and now the Bears. He didn't think any of them came close to putting him on the verge of launching his missiles, but even so, what was going on up there was pretty ugly. If there was any chance of a launch happening, even before the sub's Defcon was raised, his admiral back at Kings Bay would send him a "personal-for-commanding officer" message to let him know the world was heating up and he should brace himself.

Volonino paid closer attention to what the Russians were doing in the water than in the air. The Russian navy always staged a large sea exercise during the months of June and July. The Pentagon was following this year's maneuvers closely. Intelligence had been sent to the *Nebraska* warning that two, possibly three, Russian Akula-class subs were lurking in the Atlantic as part of the exercise. Beware.

But so far, the sonar shack had not detected any Russian subs. And as far as Volonino could tell, the Akulas had not spotted the *Nebraska*.

The boat had to postpone its return to Kings Bay to load the practice torpedoes. A message had arrived in the *Nebraska* radio shack that the USS *Maryland*, another Trident sub in the Atlantic Fleet, had plowed into a buoy as it tried to make its way through the narrow channel leading into the base. The *Maryland*'s propeller had been damaged, and remains of the buoy littered the channel, blocking other subs from sailing through it for the moment.

There was no clucking on the *Nebraska* over the *Maryland*'s misfortune, however. The crew still remembered their near collision with a buoy.

Instead of shooting torpedoes, Volonino's sub was

diverted to play rabbit for another boat practicing its hunting skills. But even those exercises were soon interrupted.

July 4 had not been special on the sub except that it was a Sunday and the workload was lighter. But Chris Wilhoite couldn't shake off this pain he felt just above his pelvis on the right side. Talk about a stomachache. It practically doubled him over to climb up and down stairs.

Wilhoite made his way back to the pharmacy to find Doc.

"I've been having this pain, like indigestion," the sonarman told Philbin, pointing to his side.

"Great," Doc said with a heave. The one-eighties were returning, he thought.

He gave Wilhoite a bottle of Mylanta, hoping the antacid would relieve the cramp. "If it gets worse, come back," Doc said as Wilhoite left.

It got worse. The next afternoon, at about one o'clock, Wilhoite limped into the pharmacy once more. The surging pain had waked him from a deep sleep the previous night.

Wilhoite looked miserable to Philbin, so he gave him a full exam. Of the ten tests he performed, one of which was a lab test, nine were positive for appendicitis. Wilhoite's white blood cells were elevated. His temperature was one hundred degrees. Philbin shook his head. This patrol was flooding with sick people. First the one-eighties, then Abercrombie's kidney stone, now a case of appendicitis, which was the most serious. Wilhoite could die if his appendix burst.

Philbin talked to no one else for the next couple of hours. He performed more tests to be absolutely certain of his diagnosis. The boat would have to turn around and race for a port. That cost money.

By five o'clock, however, Philbin had no doubt that Wilhoite's appendix was inflamed. He went to look for the skipper.

Volonino had just finished a briefing in the second-level officers study when Philbin caught him at the door.

"What now?" Volonino asked warily. Doc only buttonholed him when there was a medical problem.

"I need to talk to you in your stateroom," Philbin said somberly.

They walked forward to Volonino's room, and Volonino closed the door behind them.

"You're not going to ruin my day, are you?" Volonino asked with a wry smile as he slumped into his chair.

"We've got a case of appendicitis," Philbin said.

"What?"

"It's Wilhoite."

Volonino sat silent for a moment, taking in the news.

"This is serious, isn't it?" he finally said.

"Yes sir, it is," Doc answered. "We need to get Wilhoite out of here now."

Volonino looked at his watch, then reached for his desk and flipped open a folder that contained the top-secret operating schedule for his sub. He quickly began calculating options in his head. He was practically in the middle of the Atlantic. It would be easier for the *Nebraska* to sail back to Kings Bay than it would be to head to a European port. But they would have to move quickly.

"Draft me a message," he ordered Philbin. The Atlantic Fleet's headquarters in Norfolk had to be notified so it could approve a medevac. "We need to send one right away."

But Philbin didn't even have time to write the detailed message the sub would need to send to Norfolk on Wilhoite's condition. The sonarman had to be cared for immediately. Philbin had to rush back to the pharmacy to hook IVs to the petty officer and begin pumping him full of antibiotics and painkillers.

They decided to set up a daisy chain. While Philbin stuck IV tubes into Wilhoite's left and right arms, he began dictating the long message for Norfolk to Todd Snyder, the sonar chief, who had come back to help. Snyder relayed Philbin's sentences by phone to the radio shack at the front of the sub. On a scale of one to ten, Wilhoite's pain "has increased to seven today," the message read. "Most likely diagnosis: appendicitis."

The radio shack quickly fired off the radio transmission. Volonino wanted an answer from Norfolk fast. He was probably at least two days away from land.

Three hours later, the *Nebraska* received its reply. The medevac was approved. Fleet headquarters ordered other vessels in the ocean to clear a path for the sub so it could make a beeline to Kings Bay. It was now early Tuesday morning, July 6. Brady, the sub's navigator, calculated that at top speed the Trident could reach a tug off the Georgia coast to off-load Wilhoite by ten o'clock Wednesday morning.

Philbin hoped Wilhoite's appendix wouldn't burst before then. If his condition began to deteriorate, the corpsman had the medical books and equipment on board to operate himself. He could probably take the appendix out if he had to. But he hoped it wouldn't come to that.

Wilhoite's pain was severe; he lay on the bed in the

pharmacy, almost in a fetal position. Philbin checked the fluid level for the two IV bags, whose Ringers Lactate solution was dripping into the petty officer's veins along with two antibiotics (cefoxitin and gentamicin) plus nalbuphine hydrochloride to dull the pain.

Philbin had another worry now. He had begun the patrol with thirty-six IV bags. He had used two bags in training sessions at the outset of the voyage. Then Abercrombie had consumed twenty-five bags during the week he was hooked to an IV. Two bags were now connected to Wilhoite's arms. That left Philbin with seven bags. At the current drip rate, Doc calculated that he had enough bags to last until the day when Wilhoite was evacuated. But he was cutting it dangerously close.

All through Tuesday, Philbin kept the sonarman in the pharmacy, flooding his body with antibiotics to fight the infection in his appendix. The sub's cooks came back at different times to watch Wilhoite so the corpsman could grab a few hours sleep.

By late Tuesday night, the petty officer's vital signs had improved slightly, but his temperature remained high and the pain had intensified. Philbin couldn't even touch Wilhoite's right side without making him jump and cry out.

Doc had filled him with as much antibiotics as his body could safely take. But now he worried that it might not be enough to keep Wilhoite's appendix from bursting.

WEDNESDAY, JULY 7, 9:30 A.M.

The *Nebraska* surfaced. It had reached its rendezvous point with a tug off the Georgia coast. The sub had arrived a half hour early, which was a relief for Philbin. Wilhoite's appendix hadn't yet burst, as far as he could tell. But he was down to his last IV bag.

The men carried Wilhoite to the sub's first level near the stern, where the ladder was located to climb up to the top deck. But Wilhoite now had to make the long climb up himself.

Philbin gave him one final shot for the pain. Step by agonizing step, the sonarman crawled up each rung, the tube from the last IV bag still attached to his arm.

THURSDAY, JULY 8, 4:01 P.M.

A message from the squadron headquarters in Kings Bay came into the *Nebraska*'s radio shack. When the tug docked the day before, an ambulance had rushed Wilhoite to a Navy hospital in nearby Jacksonville, Florida, and doctors had wheeled him into the operating room early Wednesday evening.

When the surgeons opened him up, they found Wilhoite's appendix perforated. But Philbin had pumped him so full of antibiotics that the appendix's infection hadn't contaminated the rest of his body. The corpsman had "saved the patient's life," the message read. "Well done!"

20 • Love and Hate

For three days they fought. The USS *Minneapolis–St. Paul,* a sleek fast-attack submarine based out of Norfolk, had a cunning skipper who was determined to kill the *Nebraska.* For the Trident, it was the most demanding test of the patrol—the Tactical Readiness Exam, when practically every aspect of a submarine's war-fighting capability is harshly graded by senior officers from the Atlantic Fleet. If the *Nebraska* wanted to win the Battle-E—and, though the competition was intense, it was still the boat to beat for that coveted award—the crew had to score high on the Tactical Readiness Exam. All the training drills the men had run the past two months, all the simulations, all the tests by its parent squadron—they had all been the dress rehearsal for this most important exam. Now the men had to perform.

A tugboat had ferried the fleet inspectors to the Trident on July 11, the day before the battle with the *Minneapolis–St. Paul* began. For the first day, they ordered the crew to respond to simulated fires, explosions, and mechanical mishaps all over the boat. There were man-overboard drills, jammed-rudder drills, reduced-visibility drills, faulted-dive drills. One after another until the crew practically dropped from exhaustion.

Then came the *Minneapolis–St. Paul*. The testers
put an extra twist into the war game with the attack
sub. The *Minneapolis–St. Paul* would not be just the
wolf for this Tactical Readiness Exam. It would be a
hungry wolf. The fleet ordered that the attack sub's
captain also take his Tactical Readiness Exam at this
time. He would also be graded on how well he could
hunt and attack the *Nebraska*. Volonino expected no
professional courtesy from this skipper.

The two contestants met off the Florida coast. Round
one went to the *Nebraska*. The Trident simulated launching missiles, the time when it was the noisiest, and the attack sub tried to find and sink it before all the rockets
simulated leaving their tubes. The *Minneapolis–St. Paul*
failed miserably. Volonino conducted four simulated
launches and managed to sneak away each time into the
black sea before the attack sub ever got close.

Round two: Each sub took turns playing the cooperative target so the other could find it and strike. To keep
the wolf in the game—otherwise one sub would have
little hope of detecting the other—the rabbit made a
little noise, such as turning on machines or tapping a
metal pipe with a hammer. And to spice up the hunt,
the testers ordered that nothing on either sub work
properly. More fires, floods, and equipment failures.
Both boats scored high.

Round three came on the last day. *Mano a mano*
combat. No more helping the other with little noises.
Both subs would fight silently. To make it interesting,
the testers on both boats ordered each skipper to sail to
the same point in the ocean so they would be forced
into a battle. The only thing separating the two when
they met would be the different depths at which they
had been ordered to sail.

Silently, the *Nebraska* and *Minneapolis–St. Paul* slithered to their rendezvous point.

When they arrived there, the two subs crossed, so quietly that the one on top didn't know that the other was underneath it.

Volonino pointed the *Nebraska* south when it came out of the intersection. The *Minneapolis–St. Paul* sailed north.

Then, by nothing more than chance—the same dumb luck that Kinman and Habermas had had to bag their kill at the beginning of the patrol—the *Minneapolis–St. Paul*'s sonar caught a whiff of the *Nebraska* as it sailed south.

The attack sub turned right and crept into the Trident's baffles.

Within minutes, the two skippers—like chess players who can plot a half-dozen moves ahead—knew the game was over.

The *Minneapolis–St. Paul* simulated firing a torpedo. The *Nebraska* immediately began evasion maneuvers and fired back a simulated torpedo forty-five seconds later.

But there was no way for Volonino to escape the attack sub's torpedo. And the *Minneapolis–St. Paul* couldn't shake off the *Nebraska*'s counterfire.

In a real war, it would have been mutually assured destruction. The only distinction: The *Minneapolis–St. Paul* seamen would take their last breath forty-five seconds after the *Nebraska*'s seamen.

But in this war game, the attack submarine had won. The Trident had lost, by forty-five seconds.

It was a bitter defeat for Dave Volonino. The *Nebraska* had never lost a mock engagement with another sub, as far back as any of the crew could remember. During this entire patrol, the sub had kicked ass with every other war vessel it met. Volonino was a skilled tactician. During one previous exercise, he had even sneaked his submarine between two high-tech Navy destroyers that had been close together and simulated topedo attacks against both ships, without their skippers ever knowing that an enemy had sailed between their legs.

Now, to be bested by another sub—even if it was by only forty-five seconds—was a tremendous blow. Up to that point, the *Nebraska*'s men were certain they had clinched a score of "excellent" on the Tactical Readiness Exam. They had aced every other drill. But because of the forty-five-second defeat, they received an "above average" as their final exam grade. It was a good score. "Excellent" scores were rarely handed out, and only the top boats received "above average" scores. But it might not be enough to win the Battle-E, the crew now worried.

Volonino was despondent but not angry. His men had found he was like that. He would become irritated with minor foul-ups. But he would take major setbacks in stride.

"I'm proud of every one of you," he told his officers afterward. "This is the only time we've lost. It's unfortunate that it had to happen on this day. But it is the privilege of my life to be your captain."

With the channel cleared, the *Nebraska* finally was able to pull into the Kings Bay port on July 16 to load practice torpedoes for its shoot. As sailors began the ticklish job of moving the giant weapons into the submarine, Kinman sat in his stateroom, the JO jungle, cramming his last piece of clothing into a nylon bag.

He felt like a deserter. He shouldn't have. But he couldn't escape the anxiety that something was wrong. A piece of business had been left unfinished, and his men needed him.

Kinman was leaving the submarine, for good. Though this patrol wasn't yet over, his time was up in the Navy and he wanted to get out. He had served the five years the military had required of him in exchange for receiving a free education at the Naval Academy. Kinman and his wife, Jenny, wanted to have children, and he wasn't about to raise a son or daughter with three-month interruptions all along the way. Beginning next month, he would enroll in the University of Georgia and earn a master's degree in business.

Kinman nervously zipped up the nylon bag and looked around the jungle to see if he had forgotten anything. He felt as if he were leaving a womb to step out into a strange new world. The civilian world was so different from the military, and particularly from this sub. All his adult life, each day had been planned by others, from the clothes he wore to the food he ate to the manuals he read to when he woke up and when he went to bed. Now he was about to be hit with sensory overload. He could eat what he wanted, sleep when he wanted to, hop in a car and drive to the store when the urge struck him, flip on the TV and watch it for hours if he had a mind to.

The future made him almost hyperventilate. He didn't know what he wanted to do. He wanted to do something. But no one in a uniform would be there to give orders. He had strolled around the University of Georgia campus one day before the sub set sail two months ago. He was shocked. You could actually go to class wearing Nikes and sweatpants? Carry a knapsack with your books in it, something they'd never let him do at the academy? Skip class if you wanted to?

Kinman slung bags over his shoulders and began walking toward the back of the sub, where the hatch to the top deck was located. He still couldn't shake off the sense that he was abandoning the ship. There was no reason to feel guilty, he told himself. He wanted to be with Jenny. He wanted to begin grad school. He didn't want to spend the rest of his life on submarines. He had signed a contract with the United States government and he had fulfilled his part of the bargain. The government had given him a world-class education, but he had given the government a good chunk of his life.

Kinman had to stop along the passageways at different points so sailors and chiefs and different officers could give him hugs. He didn't want that, not now. He just wanted to slink off the sub and disappear. He had already talked about leaving to practically everybody on the boat during the past two months of patrol. To say goodbye now to the more than 150 members of his family would have been too much for him to bear. He had told them all during the last two months that he would keep in touch. But he knew that he would probably never see most of these guys again.

Kinman slowly climbed the rungs of the aft ladder, one hand gripping the handles on several bags. On the top deck, a Georgia July sun blazed down at him, and he squinted his eyes. It was steamy hot outside, like the tropics, and Kinman began sweating almost immediately as sailors helped him gather his bags. His skin was pale from the months underwater, and his khaki uniform sagged on him. During the patrol, he had taken a vow. He had joined the Navy weighing two hundred pounds. He would leave it at the same weight. That meant shedding the thirty-two pounds he had accumulated during his time in service.

Kinman stepped up to the gangplank that stretched from the top deck to the pier. He turned left and saluted the American flag now waving listlessly from the stern.

It still didn't feel quite right to him.

He walked silently across the gangplank and didn't look back.

With the practice torpedoes loaded, the *Nebraska* sailed south once more, arriving at the target range off Andros Island in the Bahamas on July 19. The shoot turned out to be a spectacular success. The five torpedoes the sub fired all hit five moving targets. The practice weapons actually didn't strike the target. At the last second, they were programmed to turn away so a boat could later pick up the undamaged torpedoes to be used again. Hitting five out of five was great. Most boats were happy hitting four out of five. A perfect score made it a sure bet that the *Nebraska* would now win the Battle-E, the crew was convinced.

But nobody aboard the Trident appeared in the mood to celebrate. Just the opposite. They were at each other's throats.

Hate week had begun aboard the *Nebraska*. The Trident experienced it every patrol, the week when everyone got on everyone else's nerves, the week when the men became fed up with the long workdays, when they became tired of being cooped up in the giant black pipe, looking at the same faces day in, day out.

When hate week occurred varied from patrol to patrol. But usually it came after halfway night, when the realization began to sink in that even though half the patrol was over, the other half still remained. Hate week this time had been as easy to spot as it had been during past patrols. The men had begun snapping at one another. Fights hadn't broken out, but there had been shouting matches and chests butted. Weller faced more discipline cases among his sailors.

Little problems that normally wouldn't faze the crew now turned into big squabbles. Second class petty officers who usually would pay it no mind began putting third class petty officers under them on report for being disrespectful. Chiefs grumbled about officers. Officers became testy with one another. Some complained that Boyd, the submarine's XO and Volonino's enforcer, was taking that part of his job too seriously and being too harsh with them.

Volonino struggled to keep his game face on during the dreaded week, but even he found himself growing more impatient with the men. Fatigue caused it. He was as worn out as they were. The petty bickering began to get on his nerves. Grown men were acting like children. He became irritated when they asked him the same stupid questions again and again. When he strolled by sailors on watch, he'd grit his teeth when they'd ask: "Hey, Captain, what are you doing when you get home?"

"I'm not thinking about it," he'd say with a cold stare. "And you need to pay attention to the job at hand instead of thinking about the beach."

Volonino and Weller could not treat hate week lightly, however. Not with all those nuclear weapons on board. The two leaders kept close tabs on crew morale. They knew how to get the most out of the men. But they also had to know when to let out the steam. Push too hard now and the submariners would start to balk at following orders. So the skipper and the chief of the boat did what they always did when hate week rolled around. They backed off for the moment. The schedule was lightened for a couple days.

When the men finally came out of their funk, the pace of work picked up once again. Volonino still had to squeeze more out of the crew, to get them home safely.

PART III

• • •

GOING HOME

21 • "Hard to Describe"

A boom box blared out Jefferson Airplane's "Somebody to Love." The song came from a Jacksonville radio station that played rock music from the sixties and seventies. A hot afternoon sun beat down on the black top deck of the *Nebraska* and gentle waves from the bright blue Atlantic slapped its hull. A brisk breeze blowing from the stern kept the heat from becoming stifling. But even if it had been, no one would have minded. Sailors in bathing suits and Bermudas, their bare chests as white as fish bellies, were stretched out on towels or foldout lawn chairs, desperate to soak up as many rays as possible while the sub was surfaced. Some just slept on the hard deck, while a few others read paperbacks. A bottle in the ocean drifted by, but no one could tell if it had a message inside.

The *Nebraska* was puttering along at three knots off the Florida coast. The Trident was three days away from July 30 and the end of its patrol. In three days it would finally dock at Kings Bay and the men would turn the boat over to the Gold Crew, then reunite with their families. But while it was surfaced now, Volonino had decided to have a "steel beach," as the crew called it, when the submariners could climb out of their cocoon and sunbathe.

The cooks had set up a barrel-size charcoal grill near the stern, and, as he always did during a steel beach, Volonino flipped hamburgers over sizzling coals and served them to sailors with a spatula, like the friendly neighbor at a cookout.

"I need burger boys!" Volonino yelled out when the line of sailors waiting for hamburgers thinned. He was chomping on a cigar and sported sunglasses, T-shirt, and shorts. Weller, who wore shorts as well, slapped raw patties on the grill. A metal table stood nearby groaning under pans full of baked beans, sliced cheese, onions, pickles, freshly baked hamburger buns, and pasta salad. Two plastic jugs with Kool-Aid and soda sat by the side, along with a large tin can full of chocolate chip cookies.

The boom box now blasted out the Rolling Stones. Boyd climbed up through the top hatch with papers on a clipboard for the captain to read. Volonino signed them with a free hand and returned to his burgers.

Volonino tried to fight the melancholy he felt now that the patrol was ending. He had no problem keeping up morale among his men now; it was sky-high because they knew they would soon be home. The evening before, the cooks had baked him a cake for his forty-second birthday and he had stood before it beaming as crewmen sang the University of Nebraska fight song.

But he was nearing the end of the most rewarding time he had ever had in the military, and he wanted to hang on to every moment left. He had startled the chiefs just the other morning by wandering into their quarters with nothing particular on his mind. The skipper just wanted to chat and swap jokes with them.

"Must be tough giving up your job as god in a few days," one of the chiefs joked after he left.

Everyone else on board couldn't get away from the underwater prison fast enough. To mark the last twenty-four days of the patrol, a poster of a pinup, called the "tube girl," was moved from one missile to another for the number that corresponded to the number of days left. The men were starved for contact with the rest of the world. There was a lot of catching up to do, and not just with their families. The air war in Kosovo had ended. John F. Kennedy, Jr., had died in a plane crash. Baseball's pennant race was heating up. Radioed news reports from shore had been skimpy the past two weeks, and the men were desperate for details on what had happened while they were gone.

The crew had kept count of everything. They had conducted eighty-four training drills during the patrol. Ed Martin, the engineering petty officer and one of the sub's barbers, had cut more than a hundred heads of hair. Eric Chambers, one of the sonarmen, had read ten books. Ryan Hardee had run 204 miles on the treadmill. Two of the chiefs, Sean Brown and Stacy Hines, had played 163 games of cribbage (Brown had won the match by one game). The other chiefs had memorized the words to every country music video on board. (Favorite videos for the rest of the crew: *Varsity Blues*, *Cruel Intentions*, and *A Night at the Roxbury*.) Reggie Rose had trained eight planesmen to drive the boat and persuaded two sailors to give their lives to Christ.

There were records the men were proud of. Four petty officers had received radio messages that they had been promoted to chief and had already begun their initiation rites for the exclusive club. The sub had

gone on a dieting frenzy. The chiefs alone had lost the
equivalent of one man—a total of 213 pounds. So obsessed had the crew become about shedding fat that
Rich McCloud, the sub's mess chief, had begun
polling the sailors to see if something was wrong with
the food. He had about two hundred extra pounds of
chow still crowding his freezer.

There were records the men would sooner forget.
One man faced a likely divorce when he returned
home. In the last week of patrol, fifteen crewmen had
come down with fevers and runny noses from another
virus. Doc suspected a bug sneaked aboard when the
sub docked to load the practice torpedoes.

Ryan Beasley felt as if he was catching the virus. A fitting end to this crummy patrol. He sat on a toolbox,
puffing on a cigarette, in a machinery section in the sub's
lower level where smoking was allowed. A grayish-
white cloud hung over this stuffy compartment where
sailors crowded in for their nicotine fix.

He had decided that submarining definitely
wouldn't be his life's ambition, not after these three
miserable months. The confined space hadn't bothered
him. But seeing the same faces every day, particularly
the unfriendly ones—that had become annoying. A
real character builder. Practically the only two things
he had done during the patrol were drive the boat and
crank. Cranking in the scullery was a Navy tradition
for nubs. Why, he didn't know. But it sucked, going to
work at 3:30 A.M. and washing dishes until 8 P.M. He
was amazed how once his head hit the pillow in his
rack, he fell into an instant sleep.

He was thankful that he had been allowed some time
to apprentice in the pharmacy. Doc had finally allowed

him to draw blood from another sailor's arm the other day. That had been fun. He planned to apply for training as a corpsman, try to be assigned to a Marine outfit or a surface ship. Anything but subs.

Beasley stubbed out his cigarette, grabbed the pack it had come in along with the list of names the chief of the watch had given him, and climbed up the ladder to the next level. He was one of the control center's messengers at the moment, and his job was to wake up forty people who had to start their watches. That meant walking all over the sub for the next half hour jostling men out of a sound sleep, another delightful chore.

Reaching the third level from the ladder, Beasley passed by the small trash room near the galley and caught a whiff of the foul odors from the garbage and paper refuse stacked inside it. He grunted at Shafer, who bent over inside the room, cramming trash into the compactor.

"Oh yeah, go Big Red," Shafer answered in a glum southern drawl, looking up. His red T-shirt was covered with sweat stains and streaks of dirt, and he didn't smell much better than the garbage he was packing into canisters. The cylindrical canisters were about three feet high and a foot wide and had holes in their sides to let in water. They were shot out of the sub in order to eliminate its biodegradable waste. Shafer was mildly annoyed that he had left the trash room perfectly clean at one o'clock this morning and by noon today, when he returned for his next thirteen-hour shift, it was piled up once more with crap.

But only mildly annoyed. Ridding the sub of its garbage certainly wasn't the most coveted job on board, but it was an important one, he knew. Someone had to "smash trash," and Shafer felt honored that he could help. Even after he ended his previous shift at 1 A.M., he had stayed up three more hours. He had cornered Thorson, who gave him an impromptu class on the reactor's primary and secondary coolant loops. They would ask him about the loops on his final test for his dolphins.

Scott Shafer had been astounded at the amount of work he had been able to pack into each day of the patrol. When he first looked at the ship's schedule, he had thought he would have plenty of free time during its eighteen-hour days. Six hours of watch, then twelve hours off. But the so-called time off became consumed with hours of extra jobs along with studying for his dolphins, so that in any eighteen-hour cycle he was lucky if he could squeeze out six for eating, showering, and sleeping. He had spent three solid weeks in the scullery cranking and another two weeks in the trash room. It had taken him a while to adjust his body and mind to the nonstop labor. After almost three months of it, he had been the most tired he had ever been in his life. And also the happiest.

Every spare moment on the sub, he had spent soaking up information like a sponge. He hadn't even given Sandy, the mother of his son in Texas, the address for the sub or family gram forms to write with news about four-year-old Blake. If he had received letters or messages about Blake, they would only have distracted him and made him think about home. He wanted to stay focused on his work.

The petty officers over Shafer didn't consider him the smartest sailor they had ever ordered about. He had never been good at book learning, and there were written and oral tests he would have to take to earn his dolphins. He had spent three weeks driving the boat and had taken longer than most nubs to master moving the planes. But his supervisors all agreed he was one of the hardest-working. Shafer was far ahead of the sub's other new sailors in getting items checked off on his qualification card. If he kept up the same frenetic pace ashore, he could expect to pin on his dolphins not too many weeks after the *Nebraska* pulled into port.

Shafer had been proud of what he had absorbed in the months before he boarded the sub, but it was nothing compared to what he had accomplished on the patrol. As one of the apprentice mechanics, he had learned how to operate hydraulic equipment, the sanitary tank pumps, valves on the ballast tanks, and assorted air compressors, blowers, and dryers. During torpedo drills, he rushed to the back of the boat to help at a damage control station; fire drills, he became a hose man on the number four hose; missile drills, he served as a phone talker in a lower-level machinery room where valves were flipped to keep the sub hovering at a shallow depth. The mystery and fear he had felt when he had first boarded had long dissolved. He was no longer afraid of these machines. He couldn't wreck the boat, there were so many backups to backups.

It hadn't been all rosy. Some of the petty officers were churlish, many of the nubs were immature kids, and he couldn't get more than 563 feet away from them. But never had he felt so fulfilled. There were highly technical jobs he could now perform with what he was learning on this sub. Before, he hadn't been able to see himself doing anything more than hauling wood the rest of his life. But now the Navy had opened doors he never thought would open before.

Shafer lined up the cylinders he'd packed with trash

so they could be fired off later in the evening. As soon as the *Nebraska* docked and he got off the boat, he knew exactly what he wanted to do. He'd buy himself a peanut butter cup (there'd been none on board), wash it down with a Pepsi, then find a phone booth to call Blake. He had already put in for leave next month. He'd buy a plane ticket, fly down to Texas, and spend sixteen glorious days with his son. Then he would go back to work.

FRIDAY, JULY 30, 5:40 A.M.

The warm eighty-two-degree breeze beat against their faces, so the humidity didn't feel uncomfortable. Hardee knew that the best time to be officer of the deck was early morning before sunrise when the sub was surfaced and you were on the bridge and you could enjoy the stars of the night—there were so many of them at sea—then the sunrise. The *Nebraska* had surfaced an hour and a half earlier about a hundred miles from the Georgia coast. Its time for mooring at Kings Bay had been moved up two hours, to 2 P.M., which, of course, delighted the crew.

All the men were now infected with "channel fever," the all-consuming craving to end the patrol and be home. They had spent the previous day sweeping, scrubbing, and polishing the sub from top to bottom to prepare it for being turned over to the Gold Crew. The meals had been grim the past week, because canned vegetables and frozen patties were the only foods left, but the cooks had saved Alaskan king crab legs and fried shrimp for yesterday's lunch. Then last night, the chiefs had baked pizzas and watched *Armageddon* in their quarters. They'd seen practically every other video. The sailors lay in their racks and tried to sleep, but many didn't, thinking about what they would do first when they got home.

A haze obscured the stars this morning, but it left a glowing ring around the full moon that beamed on the sea like a giant lamp. Still the ocean was dark blue, except for the frothy white wake that surrounded the sub and trailed it far behind. The only sound on the bridge came from the churning of the water as the hull plowed through it and from Hardee talking softly to Ray Chesney, who was there with him.

Chesney was getting a class from Hardee on an officer of the deck's duties on the bridge. That was the next watch station the ensign was studying. Chesney had hustled on the patrol. It made the days go by quicker, he had found, and gave him less time to dwell on how much he missed his new wife. He had already qualified as an engineering officer of the watch, an engineering duty officer, and a contact coordinator.

Hardee showed Chesney how to spot faint dots of light five to eight miles out that were "contacts"— ships near the *Nebraska*.

"Oh, I see another one," Hardee said, pointing across the starboard side of the bow.

Chesney brought binoculars up to his eyes and squinted.

"Okay, I see it," he finally said. But just barely.

Hardee grabbed the microphone and handed it to Chesney.

"Now say, 'Coordinator, bridge. I hold a second contact bearing two-nine-zero,'" Hardee prompted him.

Chesney keyed the mike and repeated the message.

"It's my last watch on the *Nebraska!* Whoaaa!"

Hardee yelled, pumping his fist in a cheer. He had been promoted to lieutenant during the patrol and would soon be transferred to a shore job. But he would miss the sunrises.

Hardee was showing Chesney the buttons to punch to read the screen for the differential Global Positioning System when he heard a different slapping sound on the waves. Submariners who have stood long lonely watches at night on the bridge learn to pick out the slightest changes in sounds from the sea. He looked up.

"There are the dolphins!" Hardee shouted, pointing to the bow.

It was a ritual the crew enjoyed every time the Trident surfaced and sailed near port. Dolphins dancing on the bow wave.

No one knew why, but the friendly underwater mammals loved to swim with the submarine when it traveled on the surface. Their favorite game was surfing on the wave that the sub's bow created as it burrowed through the water.

Four sleek gray dolphins now shot up from the bow wave and dove down, letting the force of the wave propel them along in front of the sub.

Chesney and Hardee stopped what they were doing and gazed quietly at them. It felt almost spiritual. They could see that the dolphins were having fun. One of God's gentlest creatures playing with man's most destructive war machine.

Hardee looked up at the glowing haze that had widened around the moon.

"We'll have bad weather the next three days," he predicted. He just hoped the sub would reach port before afternoon thundershowers.

At 6 A.M., Habermas, with Lieutenant Junior Grade

Rob Hill in tow, climbed up the ladder to the bridge to relieve Hardee and Chesney. Hill was apprenticing under Habermas to be an officer of the deck. A chilly shaft of air-conditioned air from the control center far below followed them and mixed with the fresh but muggy air outside. This was Habermas's last time on the bridge as well. He was leaving the Navy after this patrol. He brought up a camera to snap a few final photos. Maybe he'd show them to his grandkids one day.

Fifteen minutes later, Volonino climbed up the ladder and hoisted himself into the bridge, which was now becoming crowded.

"I thought I'd come up and see my last sunrise from here," he said, smiling and lighting up a cigar.

Don Katherman, a senior missile technician, followed Volonino up the ladder with a small electronics box that measured the temperature and humidity outside. Katherman took the weather readings to adjust the temperature in the missile tubes. Surfacing the sub affected the distribution of hot and cold air around the twenty-four missiles.

"Captain, so where are you heading from here?" Katherman asked as he watched the readouts on his box.

"Oh, nowhere important. Washington, D.C.," Volonino answered, only half joking. "So what's the condition of the *Nebraska*'s new softball team?" he asked, changing the subject.

"They claim they're good," Katherman answered.

"I expect we'll bring home the championship."

"It's ours, sir."

Volonino grabbed the mike and ordered up coffee. He looked behind him on the starboard side, where the sun would rise. The sub was pointed west, rolling along lazily toward the coast.

"Well, with this haze it's going to be a subtle sunrise," he concluded. "But sometimes they can be beautiful too."

Habermas looked back and agreed.

A sailor climbed up ten minutes later with a warm plastic pot and a stack of paper cups. He said he'd stay until the men finished their coffee (but only because he really wanted to see the sunrise). They all scrunched their shoulders to make more room.

Volonino silently sipped from a steaming cup and puffed on his cigar. Streaks of pink now began to emerge from the horizon behind them. He couldn't remember when he had last felt this contented.

The radio crackled to interrupt his thoughts.

"Bridge, control," the voice said. "Request permission to relieve the lookout."

Volonino looked behind him to the top ledge where Ashley Fuqua stood next to the periscope mast. A fire control technician, he was tall and beefy with wide eyes and sprouts of blond hair. Fuqua shook his head, no.

Volonino chuckled. Any other time of the day, the lookouts couldn't wait to climb down at the end of their shift. But never when the sun was about to rise.

Volonino grabbed the mike and keyed it. "Control, bridge. Wait one," he said, then turned around to Fuqua with a smile. "Your punishment is to remain on the bridge and watch the sunrise."

Shortly after 6:30 A.M. the sun poked up from the horizon. The view wasn't spectacular this time. Clouds and haze obscured the fireball.

But they all stopped and watched for several minutes anyway.

"The only thing bad about sunrises is they happen too quickly," Volonino finally said. Heads nodded.

"That temperature stabilized yet?" he asked, turning to Katherman.

The missile technician had long ago gotten a stable reading from his box. But he answered: "Just about." Katherman also wasn't in any hurry to climb down.

"Make sure it's good and stable," Volonino said, chuckling.

The sub was now sailing at ten knots. The white blanket of froth widened around it. The stiff breeze blew Volonino's hair back. He sipped his coffee quietly for several more minutes. The moon still shone high in the sky, but the top of the bright pink sun now nosed up over a cloud.

"I'll bet the guys on surface ships get tired of seeing this every day," Volonino said, to no one in particular. "I'll bet none of them sneak up to watch it like submariners do."

"People take it for granted," Katherman agreed.

"'Avast me swabs' and other nautical expressions," Volonino shouted dramatically.

"The one today is 'Land ho!'" Katherman corrected.

"I just started reading *Treasure Island*," Fuqua chimed in.

"Enough gazing!" Volonino finally said. He turned around to the radar screen and differential GPS screen and began quizzing Hill to see if he could spot the contacts around the sub.

At 7:05 A.M., Dave Weller, the chief of the boat, climbed up to the bridge. Katherman and the sailor carrying the coffeepot had already climbed down the ladder.

"We'll take it over for a while," Volonino said to Habermas and Hill. The two junior officers climbed down the ladder.

Weller told Fuqua he could head down as well. The COB would stand his watch for now.

Volonino and Weller wanted to be alone on the bridge for the next hour. The private moment was freighted with a special emotion for the two men. As the sub sailed west, they were ending the most intense two and a half years of their lives. Along the way, the chief and his commanding officer had become almost as close as brothers.

Volonino puffed quietly on his cigar. Weller cupped his hands around a match to light a cigarette and inhaled deeply. They began reminiscing about the patrols they had sailed on together, interrupting their memories every now and then to phone down to the control center to make course changes or report on other ships they spotted far away.

They swapped stories about the funny moments, stories about the difficult ones. There had been good and bad times, commanding the men. More than 150 high achievers operating an incredibly complex, and dangerous, weapon. It had been the biggest management challenge of both their professional lives. They had spent hours and hours huddled together during the voyages, just the two of them in Volonino's stateroom, talking about how best to lead and motivate the crew. They had shared, as well, painful moments in their personal lives, leaning on one another to cope with their divorces.

Weller would miss Volonino. He had been the best skipper the chief had ever had. Volonino was dreading pulling into the dock. If there was any way he could wangle another patrol, he'd do it in a heartbeat.

The breeze remained brisk, but with the noon sun now beating down, the day had become hotter and muggier.

The men on the bridge felt as if they were being hosed with hot steam. They had rubbed gobs of Hawaiian Tropic sunscreen on their faces and arms to keep from being burned. They all smelled like coconuts.

Mark Nowalk now took his turn apprenticing as the junior officer of the deck. Habermas stood at his side to prompt him, but the lieutenant wasn't telling him much. Volonino stood behind the two officers and did practically all of the directing. A tug had also dropped off a harbor pilot, who stood on top of the sail and would help guide the sub into port. The *Nebraska* had just passed buoy 10 and was beginning the eight-mile-long trip west in the channel that leads to the St. Marys Entrance.

Nowalk was jittery. His time on the bridge would be considerably more difficult than Chesney's. Chesney had a big ocean in which to roam. Nowalk was supposed to be assisting Volonino in sailing the sub into port along a narrow channel. Just as he had during the trip out of Kings Bay, Volonino now took over much of the steering from the bridge for the nerve-racking drive home. Nowalk's job was mostly to relay Volonino's orders to the helmsman in the control center. But Volonino constantly quizzed the ensign on the turns that should be made and occasionally let him decide the course himself. This was Nowalk's chance to shine in front of the skipper, or make a fool of himself.

Volonino clenched another glowing cigar between his teeth. Even Habermas had lit up a Tiparillo. Up to this point, the skipper had seemed to Nowalk to be fairly relaxed on the bridge. Not worried about being covert any longer, he'd ordered that the sub's number—739—be slapped up on the side of the sail "to let everyone know the *Nebraska* is coming home." But as the sub entered the first eight-mile leg, the commander became tense and more focused on keeping the boat centered in the channel.

Nowalk's muscles tightened. His stomach churned. He gripped the microphone like a hand grenade, instantly ready to relay the skipper's orders.

"Passing buoy ten," said Eric Chambers, who stood behind the officers on the top of the sail next to the periscope that stuck up. He peered through his binoculars. The sonarman was the lookout for the trip in, just as he had been for the trip out.

Volonino eyed down the path ahead of the sub.

"Okay, we've just entered the channel and we're beginning the eight-mile leg," he lectured Nowalk. To give the sub a little room from the other buoy on the left, he ordered, "Let's come right to steady course two-six-eight-point-five."

Nowalk tensed and squeezed the mike. "Helmbridgecomerightsteadycoursetwosixeightpointfive!" he blurted into the microphone like a machine gun rattling off rounds.

Volonino patted Nowalk on the shoulder to calm him down.

"Okay, you don't have to shout it to control," he said soothingly. "You'll scare them to death down there."

Nowalk blushed and bit his lip. Idiot move number one. Even the slightest criticism from the skipper crushed him. Nowalk worshiped Volonino, his only regret being that he would serve just one patrol under him. Then he would have to prove himself again to a new commanding officer.

Nowalk had tried to keep his eyes and ears wide open during the patrol and vacuum as much information as he could. Being an officer certainly had its advantages over his previous life as a sailor—extra perks, higher pay, more responsibility. In the enlisted ranks he had specialized in one job. As an officer he was an overseer, and was expected to know practically everything about the ship. He still was getting used to being served by waiters in the wardroom and not having to clean his dishes afterward. Now if he could just get out of the crew quarters—he despised sleeping there—and move into an officer's stateroom. He hoped to do that the next patrol.

Probably the fact that Nowalk had experience as an enlisted man kept him from getting picked on as much as Chesney in the wardroom. He had also stayed slightly ahead of Chesney in qualifying at different watch stations. He expected to pin on his dolphins early in the next patrol.

The sub now was sailing at a brisk fifteen knots, its bow covered almost to the sail in the glassy calm water. Far ahead, at the end of the eight-mile leg, Nowalk could see through his binoculars the St. Marys Entrance, the part of the channel that ran between Amelia Island in the south and Cumberland Island to the north. Fortunately, during the trip to that entrance, he was sailing against an ebb current—an outbound flow of water—which meant the ship responded better to rudder movements because the water traveled faster over its surface. But he had to watch for crosscurrents, which pushed the sub south off the center track of the channel.

He also had to be on the lookout for speedboats racing dangerously near the sub. Usually their operators were rednecks gunning their fishing boats across the sub's path for the hell of it or sightseers in pleasure craft sailing too close to gawk.

"Passing buoys eighteen and nineteen," Chambers yelled from his lookout station.

Two buoys down, however, Volonino could see he would have trouble not with a civilian vessel, but with one of the Navy's own small craft. A security speed-boat that had been ahead of the *Nebraska* to shoo away intruders was now dead in the water and drifting to the middle of the channel.

Volonino quickly ordered the control room to slow down the sub, almost to a stop. Otherwise the Trident would run over the speedboat.

The *Nebraska* began decelerating. But now Volonino worried about another problem. The current swirled at this particular point in the channel. If the sub slowed too much, the rudder wouldn't bite into the water and the current would push the sub out of the channel.

The speedboat's engine finally kicked in and the craft raced away from the center of the channel.

Voloninio began spitting out orders quickly for rud-der turns and revving up the *Nebraska*'s engines. But Nowalk couldn't keep up with relaying the instructions quickly on his phone to the control center.

The swirls were still working against the sub.

"We're drifting off course here," Volonino said, his voice rising in irritation. "We've got to get back on course." He didn't want another problem such as they'd had sailing out of Kings Bay when the sub reached a dead stop in the channel and it became diffi-cult to line up and move the vessel again.

Nowalk grew more flustered and stuttered out the commands.

"You've got to move on this, Mr. Nowalk," Volonino

snapped. "It doesn't take long for this ship to get into trouble."

"Yes sir," Nowalk said meekly. Mistake two.

Volonino finally got the Trident speeded up again so its rudder bit into the water. But he was still nervous from his brush with another calamity, and what he saw ahead left him more edgy.

The Navy security boat had moved ahead of the *Nebraska,* and again to the center of the channel, in order to escort the sub. If its engine had conked out once, it could stall in the middle of the channel again and Volonino would have to repeat the nightmare of stopping and being shoved out of alignment by the currents. The speedboat's operator should have had enough sense to realize that. Had everybody in this Navy gone brain-dead this morning?

"Tell that boat to stay out of the fricking channel and don't creep in front of me again!" he shouted, turning to the harbor pilot. "Unless it has a death wish!"

The harbor pilot quickly brought his walkie-talkie to his lips and radioed to the speedboat to exit the channel and instead sail beside the giant sub.

The *Nebraska* was now nearing the end of the eight-mile leg and the men on the bridge could see Fort Clinch on the tip of Amelia Island to their left. Volonino lifted himself up to sit on the top of the sail so he had a better view. The sub now had to make a series of three quick turns to the right to come around Cumberland Island to the north and then sail up Cumberland Sound to the dock. But the three right turns could be treacherous. The sub was traveling at twelve knots, or about four hundred yards a minute, and each of the three legs in the swing to the right was only three hundred yards long, or less. What's more, there was a whirlpool of crosscurrents battering the boat during these turns, which forced Volonino to alternate crabbing right or left to stay on the center line.

Volonino had to think fast. This part of the voyage could be treacherous. He had faced the swirling currents in these turns many times before, but the conditions were never predictable, so he had to stay alert and make course corrections quickly.

And Nowalk had to relay the course corrections quickly and accurately.

Brady, who was down in the control center plotting the trip, radioed to the bridge that the sub needed to adjust its course so it would be positioned properly for the first right turn.

"Recommend come left to two-six-five," came Brady's voice over the bridge's speaker.

"Come right," Nowalk repeated quickly. Wrong.

"Come left," Volonino shouted in the next instant.

"Correction, come left," Nowalk said into the microphone.

"Think first! Then speak!" Volonino snapped.

"Yes, sir," Nowalk said softly. Mistake three. This training session had unraveled for him.

"I have an excellent fix," Brady's voice on the speaker continued, as the boat neared the first right turn six hundred yards away. "Next course is to the right two-nine-four."

"Navigator, bridge, aye," Nowalk answered on his microphone, still spitting out the words too quickly.

Several minutes later, Brady radioed up that the sub should begin the first right turn and head on the course of 294 degrees.

Volonino quickly approved it. Chambers looked back to make sure the rudder turned as ordered.

"Rudder is right, sir," he said. Fort Clinch passed by on the left.

It took less than ten minutes to complete the three right turns. The time raced by in a blur for Nowalk as he relayed a stream of orders from Volonino to crab right or left to fight the swirling currents and keep the sub in the center of the channel.

Finally, however, the sub was home free. It pointed north on a course of four degrees. Cumberland Island stretched along its right. On the left sat the marshes and tiny inlets in front of the Georgia coast. Nowalk could see the small Drum Point Island ahead and just to the right of the channel. The deep-draft boat squatted as it speeded up, which gave it as little as five feet clearance from the bottom of the channel. The trick now was to keep the Trident centered in the two-hundred-yard-wide channel. The deepest point of the channel was at its center. But many of the buoys had moved out of line, so Nowalk couldn't always count on being in the center if the boat was exactly between the buoys on both sides.

"Passing buoy four-one," Chambers announced from his lookout post. The officers on the bridge watched silently as the bobbing hunk of green metal drifted by on the left. That was the buoy the *Nebraska* had almost tapped on the way out.

Volonino scanned what was ahead of him through his binoculars. Brady phoned up that his navigators had an excellent fix on the sub's position and it was perfectly on course. As the sub neared the northern tip of Drum Point Island on the right and prepared to veer to the left once more, the giant magnetic silencing facility came into view off the port bow. Volonino dropped his binoculars down from his eyes and finally relaxed.

The sub slowed to eight knots. Two tugs waited ahead to escort the Trident on the last mile to the dock.

Nowalk choked as he relayed a course order to the control center.

Volonino now chuckled. He knew his ensign had just been through one of the most miserable days of his short career as an officer.

"You swallow some air," Volonino said, patting Nowalk on the back.

"Yes sir," Nowalk said with a weak smile.

They stood silently for several more minutes as the two tugs pulled up on each side of the Trident to guide it on the final legs, past the explosive-handling wharves on the left. Crewmen had already climbed out onto the topside and begun pounding white cleats into place with rubber hammers. Then they grabbed thick lines tossed from the tugs and tied them to the cleats. The sub's engine powered down and the tugs began pulling the boat slowly to the pier.

"How do you feel, sir?" the ensign finally asked his skipper.

"Hard to describe," Volonino said, just looking ahead. "Hard to describe."

It was almost two o'clock in the afternoon, and there was no breeze now. The oppressively humid heat had already soared past ninety degrees. A clutch of women, many straddling babies or holding children's hands, stood at wharf number three waving fans at their sweaty faces, hoping for some relief from the baking oven. A sign hanging from the wharf's office building read WELCOME HOME NEBRASKA.

As the Trident neared the dock, the tugs began nudging its hull to turn the giant sub around so it could be nestled against the pier with its bow pointed south. Volonino ordered rounds of horn blasts to entertain the families standing on the dock and annoy the crews of the other subs already moored. The wives covered their ears during the roars and the children giggled and squealed.

Nowalk peered through his binoculars to see if he could spot his four-year-old daughter in the crowd. He saw a little girl in a red dress and green hat, but he didn't know if it was Mikaela.

Thorson, who was in the control center below, had better luck. He had hopped up onto the conn, pressed his face to one of the periscope's eyepieces, and begun swiveling it around until he saw Kyung standing on the dock in shorts waving to the boat.

Dockworkers began throwing heavy lines to sailors on the sub's top deck, who hitched them to cleats to secure the sub snugly to the pier.

"Stand by to shift colors," Volonino ordered. Chambers stood up and unfastened the line holding up the American flag.

"Shift colors," Volonino commanded.

Chambers lowered the flag. The sailors on the top deck stood at attention. While in port, the flag would hang from a pole at the sub's stern.

Volonino gathered up his cigar case and charts.

"All right, finish tying up and get the bridge broken down," he said to Habermas, lifting up the floor grate to climb the ladder down to the control center.

"Aye, aye, sir," Habermas answered.

"Good work, guys," Volonino said crisply and climbed down the ladder.

Thorson stood at the back end of the missile compartment on its top level, pacing in circles nervously. Chief Lewis came up to him with a clipboard and papers he had to sign.

Thorson scrawled his name on the forms, not paying much attention to what he was signing, and continued looking up the long ladder to the top hatch, where he caught a glimpse of blue sky every now and then.

"Anything else?" the lieutenant junior grade asked, turning back to Lewis.

"Naw, that's it for now," the chief answered with a smile. Thorson looked to him like a kid waiting to open his presents at Christmas.

Volonino's voice came over the sub's loudspeaker. "I hate to keep bringing this topic up. But it keeps coming up. First, congratulations on another fine job. But a shipmate on the USS *Kentucky* was killed in a motorcycle accident. Don't let this happen to us. Be careful when you're behind the wheel."

Thorson barely listened to the announcement. His mind was on Kyung. Two weeks ago, when a tug had pulled alongside the sub, there had been a letter from Kyung in the mailbag that had been passed over. She had gone out and bought a new car by herself. A 1999 Honda Accord. He cringed when he read it. He had planned to research prices when he returned, then begin visiting lots, but Kyung had jumped the gun on him. God, he hoped the dealer hadn't ripped her off.

But he knew he'd have to ooh and aaah over the purchase when he first saw her and tell her how proud he was that she'd bought the car on her own. After all, *she* was supposed to be making the decisions when he was away. The worst thing he could do was second-guess her. Maybe later he'd ask her what the monthly payments were. And when she told him, he couldn't let her see him gritting his teeth.

Many of the officers and chiefs had told their wives not to come down to the pier right away. It usually took several hours after the boat docked before the men finished chores and could walk off. No use having the women standing around sweltering during that time.

Thorson, however, had asked Kyung to be there when the boat pulled up. He had duty on the sub tomorrow, which meant he'd be spending a drearily long day pushing paperwork for the transfer of control from the Blue Crew to the Gold Crew. But he was free to leave today, as long as he showed up first thing in the morning.

If only he could get off the damn boat. A line of sailors stretched from the top deck, down the ladder, and through the passageway of the missile compartment, passing up hundreds of black plastic bags of trash that hadn't been compacted and shot out of the sub during the last week.

Thorson quietly fidgeted. Enough bags to fill a city dump, he grumbled to himself.

Finally, after thirty minutes, the trash brigade had finished.

Sailors quickly began lining up at the bottom of the ladder to climb up and get off the boat.

Thorson found a spot to slip into the long line and waited his turn.

A minute later, he emerged out of the top hatch. The heat from outside hit him like a blast wave, and he adjusted his eyes to the near-blinding sun.

With a spring now in his step, he hopped up to the lip of the gangplank, turned left to salute the flag, and stepped quickly down the wooden plank.

Winding his way through the crowd of submariners hugging women and children, he finally spotted Kyung down the pier near the sub's bow.

She felt hot and sticky and thought she looked a wreck after wilting for more than an hour in the sun. But Thorson thought she looked absolutely beautiful.

He sprinted up to her. He kissed her, then wrapped his arms around her. He kissed her again and went back to hugging her. He didn't want to ever let her go.

• Postscript

It was a bright sunny morning that would soon turn into another broiling summer day in Georgia. The Kings Bay Naval Base's open-air brick pavilion honoring World War II submariners was filled with women in pretty dresses and men in business suits or sport shirts and slacks, all sitting on rows of folding metal chairs. Some of the men had *Nebraska* blue baseball caps on their heads and cameras hanging from straps around their necks. Other men—veterans of the submarine service now gray-haired and potbellied—wore bush hats with bright-colored plumes and vests covered with old military insignia.

On the right behind the pavilion, a small Navy band played marching music. The men of the USS *Nebraska*'s Blue Crew stood in formation to the left in their formal summer uniforms with medals hanging from their chests. The sub's officers, resplendent in dress white tunics with gleaming ceremonial swords strapped to their waists, had escorted the women to their seats.

It was a festive day, a time to celebrate and enjoy the accolades. But Volonino had dreaded its arrival: his change of command ceremony, when he formally turned the Trident over to its next skipper, David Dittmer, a Navy commander from New Jersey, who was eager to take his turn at captaining the sub. The people important to Volonino and the *Nebraska* had turned out for the passing of leadership. His father, Dominick, sat beaming in the front row (his mother was recovering from heart surgery and couldn't make the long trip south). The four-star admiral in charge of the U.S. Strategic Command was there, along with senior Navy officers from the base, the former senator from Nebraska James Exon, a gaggle of Nebraska state officials, and sixty members of the Big Red Sub Club from Nebraska.

The band played "Ruffles and Flourishes." The civilians jumped as two mini-cannons blasted a seventeen-gun salute. Ed Martin, the sub's petty officer and minister, delivered a prayer. There were laudatory speeches from senior officers. An admiral pinned another Meritorious Service Medal on Volonino's chest. Then Volonino stood at the podium, which was draped in red, white, and blue bunting, and unfolded the notes to the speech he had spent hours composing.

He looked out at the audience, saying nothing for a few seconds. Change of command ceremonies could be emotional for departing skippers. Some became teary-eyed. Volonino did not want that to happen to him. But he was choked up inside. The crew had become his family, and this was worse than losing your family. He was giving it away.

"I want to tell you what I love about this crew and why I will miss them so much," he began, looking first at his men to his right and then to the civilians seated in front of him. "To me, the sailors of the USS *Nebraska* exemplify America at its very best. Believe me, the values of honor, courage, and commitment are alive and very well and residing in these officers, chiefs, and enlisted men."

He went on to recite a brief history of submarines and to explain how important they were in modern warfare and how critical the Tridents of today were for deterrence, even with the Cold War over. Then he spoke from the heart.

"A former commanding officer of mine once told me the two most difficult tasks the Navy would ever ask of me were to command a nuclear warship and then later to relinquish that command," he said. "The first task was about as hard as I expected it to be. But even though I've tried to prepare myself for this day, I'm finding the second task a bit more than I had imagined."

Volonino paused and took a deep breath. He then turned to his men again and thanked them, naming many of the officers and chiefs to whom he had grown close the past two years. "Before relinquishing command, I have some final instructions for you," he said at last.

The crew snapped to attention.

"Remember to hold the flag high, let them see your pride," he continued. "Remember that I believe in you. You must carry on your mission without me."

He read his orders transferring him to Washington. "Commander Dittmer, I am ready to be relieved," Volonino said and stepped away from the podium.

Dittmer stood up and read his orders, then turned to Volonino.

"Commander Volonino, I relieve you, sir," he said, standing proudly at attention.

"I stand relieved, sir," Volonino replied.

And that was it. The hour the ceremony took flew by for him. Dittmer now commanded the men and their loyalty. Volonino had to break his ties immediately and move on. Phone calls back to the crew, even just to reminisce, were considered bad form. He had to let Dittmer now be the skipper.

A sailor blew a boatswain's pipe, which let out a high whine. A ship's bell clanged four times. Volonino walked down a red carpet from the podium to a waiting van.

The good news came on January 1, 2000. Volonino's men had won the Battle-E for 1999. The award, for them, tasted even sweeter because the squadron brass, for the first time, had decided to present it only to the Blue Crew.

• Acknowledgments

Authors of nonfiction can never claim sole credit for the books they write. Many people always have a hand in the work. Never has that been more the case than with *Big Red*. To write a book about a Trident nuclear submarine a lot of people have to be involved in the project—to grant the author access to these deadly war machines, to assist him in understanding the submarine's complex technology and the men who master it, to help him accurately describe this strange world to outsiders.

The U.S. Navy submarine community is secretive by nature. The service doesn't warm quickly to outsiders. I had written several books on the armed forces that had been well received in military circles, but that didn't give me a free pass among the submariners. I was eventually granted more access to a Trident than any journalist had ever had, but it didn't happen overnight.

I am indebted to a number of naval officers who aided me along the way. Now retired Rear Admiral Kendell Pease was the first to enthusiastically embrace my project when he was Chief of Naval Information, and to begin to open doors for me. Lieutenant Commander Mark McCaffrey of the public affairs staff for

the Naval Submarine Forces Atlantic helped me decipher Trident training and sea tour schedules to begin
the research. I want to thank Rear Admiral Joseph
Henry, commander of Submarine Group Ten, who
granted me the access and the extended sea time
aboard one of his Tridents and who spent many hours
giving me the "big picture" of this force. Commander
Terry Evans, the public affairs officer at the Kings Bay
Naval Base, deftly overcame a number of administrative and logistical hurdles to satisfy the many requests
I threw at him.

But ultimately, it was the Blue Crew of the USS *Nebraska* who made this book possible. I couldn't have
asked for more cooperation from every man aboard.
They were busy people at sea and on land, but they always took time out from hectic schedules to answer
my countless questions. I was the stranger, but they
graciously accepted me as a member of their submarine family. I spent five weeks with the crew while it
was training at the Kings Bay Naval Submarine Base,
getting to know the submariners and learning the complex technology of their vessel. To tell the story of a
three-month patrol of the *Nebraska,* I spent a total of
about three weeks underwater with the sub at different
times. I also used crew logs and diaries to chronicle
other parts of the patrol. By the end of my research I
had conducted in-depth interviews with 106 of the 162
men in the *Nebraska*'s Blue Crew.

I particularly want to thank Commander Dave
Volonino, the captain of the *Nebraska.* He spent literally days sitting through interviews with me and pa-
tiently explaining the *Nebraska*'s operation and how he
commanded the sub. I suspect that not many Trident

skippers would feel secure and confident enough to allow a reporter to roam their ship and interview whomever he pleased without the boss looking over his shoulder. Volonino did, and soon I learned why. He was a good sub captain, proud of his boat, whom the men respected. The officers, chiefs, and sailors under him became my tutors. I also want to thank Chiefs Jeff Spooner and Todd Snyder, who served as my guides during different voyages; Chief of the Boat Dave Weller, who counseled me on the enlisted ranks; Lieutenant Brent Kinman, and Lieutenant Steve Habermas, who provided invaluable technical advice; and Petty Officer Third Class Ben Dykes, who shot hundreds of photographs for me.

I was blessed with two fine editors for this work at HarperCollins: Paul MacCarthy, who began the project with me, and Henry Ferris, who skillfully took it to completion. My agent, Kristine Dahl, as always, energetically supported my work. At *Time* magazine, I want to thank Michael Duffy, who couldn't have been more encouraging; Mark Thompson, who reviewed the manuscript; and Ann Moffett, who cheerfully tracked down for me news stories on the military. James C. Hay, editor of *The Submarine Review*, was helpful with historical articles on the Trident. Bruce Blair of the Center for Defense Information guided me through the complexities of nuclear command and control. Kathy Davidson, with the Undersea and Hyberbaric Medical Society, dug up stacks of studies for me on submariner physiology. And Emma Hornby came through with a critical piece of research.

Finally, I want to thank my loving wife, Judy. She

kept the house together during my year of writing, research, and many weeks at sea. And, as always, she
was my best editor. It is to her mother that this book is
dedicated—from both of us.

• Source Notes

Prologue

The prologue is based on interviews conducted with senior officers at the U.S. Naval Submarine Base Kings Bay as well as with crewmen aboard the USS *Nebraska*. Interviews also were conducted with Robert S. Norris and Matthew McKinzie of the Natural Resources Defense Council.

Publications consulted included *The Effects of Nuclear Weapons*, Samuel Glasstone and Philip J. Dolan, eds. (Washington, D.C.: U.S. Government Printing Office, 1977); *Life After Nuclear War: The Economic and Social Impacts of Nuclear Attacks on the United States*, Arthur M. Katz (Cambridge, Mass.: Ballinger, 1982); *The Cold and the Dark*, Paul R. Ehrlich, Carl Sagan, et al. (New York: Norton, 1984); "Chilly Scenes of Nuclear Winter," Len Ackland, *New York Times*, Jan. 6, 1991, p. 7; "Nuclear Winter Theorists Pull Back," Malcolm W. Browne, *New York Times*, Jan. 23, 1990, p. 1; "The Lifecycle Costs of Nuclear Forces: A Preliminary Assessment," Steven M. Kosiak, Defense Budget Project, October 1994; "Questions Raised on Trident Subs," Walter Pincus, *Washington Post*, Jan. 3, 1999, p. A22; "Sea Change for Subs: Fleet Without a Mission as Cold War Fades," Patrick J. Sloyan, *Newsday*, April 22, 1998, p. A6; "Maneuvers Show Russian Reliance on Nuclear Arms," Michael R. Gordon, *New York Times,* July 10, 1999, p. 1; "The Typhoon Saga Ends," Norman Friedman, *Proceedings*, Feb. 1999, p. 91; "Invitation to Nuclear Disaster," Michael Krepon, *Washington Post*, May 25, 1999, p. A15; "Admiral Cites Crew Intimidation as Reason Trident Sub Skipper Was Removed," Bradley Graham, *Washington Post*, Sept. 18, 1997; testimony by Rear Admiral. L. E. Jacoby (Director of Naval Intelligence), Vice Admiral F. P. Giambastiani (Commander Submarine Force U.S. Atlantic Fleet) and Rear Admiral Malcom I. Fages (Submarine Warfare Division, Office of the Chief of Naval Operations N87) before the Sea Power Subcommittee of the Senate Armed Services Committee on Submarine Warfare in the 21st Century, April 13, 1999; "USS *Nebraska* Welcome Aboard," handbook, D. M. Volonino, Commanding Officer; "The Origins of Overkill," David Alan Rosenberg, *International Security*, Spring 1983, Vol. 7, No. 4, pp. 3-71; and "Why the U.S. Navy Went for Hard-Target Counterforce in Trident II," *International Security*, Fall 1990, Vol. 15, No. 2, pp. 147–190.

The websites consulted included U.S. Naval Base Kings Bay, www.subasekb.navy.mil; SUBLANT, www.norfolk.navy.mil; and Navy Chief of Information, www.chinfo.navy.mil. Also consulted: "Nuclear Notebook: U.S. Strategic Nuclear Forces, End of 1998," Robert S. Norris and William M. Arkin, Jan./Feb. 1999, Vol. 55, No. 1, www.bullatomsci.org; "Table of U.S. Strategic Nuclear Forces, end of 1996/ Notes," The NRDC Nuclear Program: Nuclear Data, www.nrdc.org.

Chapters 1 and 2

These chapters are based on interviews conducted aboard the USS *Nebraska* and at the Kings Bay Naval Base with Commander Dave Volonino, Ensign Ray Chesney, Lieutenant Junior Grade Chad Thorson, Lieutenant Al Brady, Chief of the Boat David Weller, Lieutenant Brent Kinman, Lieutenant Steve Habermas, Senior Chief Tom Standley, Petty Officer First Class Brett Segura, Petty Officer Second

Class Matt Suzor, Lieutenant Ryan Hardee, Chief Todd Snyder, Lieutenant Fred Freeland, Captain Jerry Hunnicutt, Petty Officer First Class Ed Martin, Petty Officer Third Class Eric Chambers, Chief Jeffrey Spooner, Petty Officer First Class Ed Stammer, Petty Officer Second Class Todd Jenkins, Petty Officer First Class Tony Holmes, and Lieutenant Junior Grade Dave Bush, plus senior officers at the Kings Bay Naval Base.

Publications consulted included *The Navy Times Book of Submarines: A Political, Social and Military History*, Brayton Harris (New York: Berkley Books, 1997); *Submarine Design and Development*, Norman Friedman (Annapolis, Md.: Naval Institute Press, 1984); *Submarine: A Guided Tour Inside a Nuclear Warship*, Tom Clancy (New York: Berkley Books, 1993); and "Peace Plan Reaction Is Muted in Belgrade; News Media Reflect Leaders' Cautions," Daniel Williams, *Washington Post*, May 7, 1999, p. A34.

The websites consulted included U.S. Naval Base Kings Bay, www.subasekb.navy.mil; and Navy Chief of Information, www.chinfo.navy.mil. The author also observed the submarine piloting and navigation trainer at the Trident Training Facility, Kings Bay Naval Base.

Chapter 3

This chapter is based on interviews aboard the USS *Nebraska* and at the Kings Bay Naval Base with Ensign Mark Nowalk, Lieutenant Brent Kinman, Chief Jeffrey Spooner, Lieutenant Fred Freeland, Chief Shawn Brown, Chief Dan Montgomery, Seaman Second Class Scott Shafer, and other members of the *Nebraska* crew. Sandy Gertz Sellers was also interviewed.

Publications consulted included "The CPO Creed"; USS *Nebraska* SSBN-739, Commissioning Booklet, July 10, 1993; and *Diving and Subaquatic Medicine*, Carl Edmonds, et al. (Boston: Butterworth Heinemann, 1992).

Chapter 4

This chapter is based on interviews conducted aboard the USS *Nebraska* and at the Kings Bay Naval Base and the Trident Training Facility with Commander Dave Volonino, Lieutenant Ryan Hardee, Chief Jeffrey Spooner, Chief Stacy Hines, Lieutenant Brent Kinman, Chief Shawn Brown, Master Chief Marc Churchwell, Lieutenant Commander Harry Ganteaume, Petty Officer Third Class Keith Larson, and Petty Officer Third Class Jason Bush, along with other crew members from the *Nebraska* and officers from Submarine Squadron 16. Ship's manuals for the SSBN-726 Class submarine on "Diving Procedures" also were consulted.

The author also observed the dive trainer at the Trident Training Facility, Kings Bay Naval Base.

Chapter 5

This chapter is based on interviews conducted aboard the USS *Nebraska* and at the Kings Bay Naval Base with Ensign Raymond Chesney, Ensign Mark Nowalk, Lieutenant Brent Kinman, Lieutenant Stephen Habermas, Lieutenant Junior Grade Noah Hayward, Chief of the Boat David Weller, Commander Dave Volonino, Lieutenant Commander Duane Ashton, Senior Chief Stephen Dille, Petty Officer Third Class Jason Dillon, Petty Officer First Class Christopher Cornelius, Petty Officer First Class Kirby Shuler, Petty Officer First Class Edward Martin, Petty Officer First Class Richard Garvin, and other officers and enlisted men on the submarine.

Chapter 6

This chapter is based on interviews conducted aboard the USS *Nebraska* and at the Kings Bay Naval Base with Lieutenant Fred Freeland, Chief Todd Snyder, Petty Officer Second Class Jason Barrass, Petty Officer Third Class Ben Dykes, Petty Officer First Class Roger Humphrey, Petty Officer Second Class Chris Wilhoite, Petty Officer Third Class

Eric Chambers, Petty Officer First Class Michael Fregoe, Petty Officer Second Class Russell Avacato, Petty Officer Third Class Jason Rodgers, Petty Officer Third Class Matt Douvres, and Petty Officer Third Class James Johnson.

Chapter 7

This chapter is based on interviews conducted aboard the USS *Nebraska* and at the Kings Bay Naval Base with Lieutenant Commander Alan Boyd, Commander Dave Volonino, Chief Marvin Abercrombie, Chief Bob Lewis, Chief Shawn Brown, Lieutenant Commander Duane Ashton, Lieutenant Junior Grade Noah Hayward, Petty Officer Second Class David Smith, Petty Officer Second Class Don Lee, Lieutenant Brent Kinman, Senior Chief Stephen Dille, Petty Officer Third Class Gregory Murphy, Ensign Ray Chesney, Master Chief Marc Churchwell, and Chief Bob Philbin.

Publications consulted included "Preventive Aspects of Submarine Medicine," K. E. Schaefer, ed., *Undersea Medical Research: Journal of the Undersea Medical Society*, Vol. 6 Supplement, Dec. 1979, pp. S7–S244; "Work and Rest on Nuclear Submarines," Arthur N. Beare et al., Report No. 946, Naval Submarine Medical Research Laboratory, Groton, Conn., Jan. 1981, pp. 1–19; "Nutrition Education and Diet Modification Aboard Submarines," Caron L. Shake and Christine L. Schlichting, Report No. 1201, Naval Submarine Medical Research Laboratory, Groton, Conn., June 27, 1996, pp. 1–14; "The Relationship of Job Performance to Physical Fitness and Its Application to U.S. Navy Submarines," Brad L. Bennett and Kenneth R. Bondi, Report No. 962, Naval Submarine Medical Research Laboratory, Groton, Conn., Sept. 29, 1981, pp. 1–14; "Relationships Between Perceived Health and Psychological Factors Among Submarine Personnel: Endocrine and Immunological Effects," Ragnar J. Vaernes, et al., *Proceedings of the XIIIth Annual Meeting of the European Undersea Biomedical Society*, Palermo, Italy, Sept. 9–12, 1987; and "Physical Fitness in a Submarine Community as Determined by the U.S. Navy Health and Physical Readiness Test," B. L. Bennett, *Aviation, Space, and Environmental Medicine*, May 1987, pp. 444–450.

The author also received three days of instruction in the flooding and fire trainers in the Trident Training Facility at the Kings Bay Naval Base.

Chapter 8

This chapter is based on interviews conducted aboard the USS *Nebraska* and at the Kings Bay Naval Base with Commander Dave Volonino, Lieutenant Commander Duane Ashton, Master Chief David Weller, Lieutenant Fred Freeland, Seaman Scott Shafer, Lieutenant Brent Kinman, Chief Jeff Spooner, Petty Officer First Class Richard Garvin, Chief Todd Snyder, Lieutenant Commander Harry Ganteaume, Petty Officer First Class Mike Fregoe, Petty Officer Second Class Jason Barrass, Petty Officer Second Class Chris Wilhoite, and Petty Officer Third Class Ben Dykes.

Publications consulted included "That Sinking Feeling," Mark Thompson, *TIME*, Feb. 8, 1999, pp. 34–35; "Bring Back ASW—Now!" Art Doney and Steve Deal, *Proceedings*, March 1999, pp. 102–104; Rear Admiral Malcom I. Fages, Director, Submarine Warfare Division Office of the Chief of Naval Operations, testimony before the Senate Armed Services Committee, Seapower Subcommittee, April 13, 1999; Rear Admiral L. E. Jacoby, Director of Naval Intelligence, testimony before the Senate Armed Services Committee, Seapower Subcommittee, April 13, 1999; and Vice Admiral E. P. Giambastiani, Commander Submarine Force U.S. Atlantic Fleet, testimony before the Senate Armed Services Committee, Seapower Subcommittee, April 13, 1999.

The author also observed the attack center trainer at the Trident Training Facility, Kings Bay Naval Base.

Chapter 9

This chapter is based on interviews conducted aboard the USS *Nebraska* and at the Kings Bay Naval Base with Petty Officer Third Class Frank Levering, Lieutenant Brent Kinman, Lieutenant Steve Habermas, Commander Dave Volonino, Lieutenant Junior Grade Ryan Hardee, Lieutenant Junior Grade Noah Hayward, Petty Officer Second Class Jason Duff, Chief Richard McCloud, Lieutenant Joe Davis, and Petty Officer First Class Scott Humphrey.

Also consulted were "USS *Nebraska* SSBN-739 Blue Ship's Menu," Refit 18 Cycle #1 and #2, and Patrol 18 Cycle #1.

Chapter 10

The periscope section of this chapter is based on interviews conducted aboard the USS *Nebraska* and at the Kings Bay Naval Base with Petty Officer Second Class Reginald Rose, Chief Shawn Brown, Chief Jeffrey Spooner, Petty Officer Third Class Quentin Albea, Petty Officer Third Class Edward Verbilla, Lieutenant Al Brady, and Commander Dave Volonino. Also consulted were ship's manuals for surfacing an SSBN-726–class Trident. During the patrol, the author also operated the inboard plane for thirty minutes under instruction.

The radio shack section of this chapter is based on interviews conducted aboard the USS *Nebraska* and at the Kings Bay Naval Base with Petty Officer First Class Eric Liebrich, Petty Officer Third Class Tom Horner, Chief Dan Montgomery, and Petty Officer Third Class Gregory Migliore. The author also reviewed unclassified message traffic and news reports aboard the *Nebraska* and observed training in the simulated radio room at the Kings Bay Naval Base's Trident Training Facility. Also consulted were *The Logic of Accidental Nuclear War*, Bruce G. Blair (Washington, D.C.: Brookings Institution, 1993) and *Managing Nuclear Operations*, Ashton B. Carter, et al. (Washington, D.C.: Brookings Institution, 1987).

Chapter 11

This chapter is based on interviews conducted aboard the USS *Nebraska* and at the Kings Bay Naval Base with Commander Dave Volonino, Lieutenant Junior Grade Chad Thorson, Lieutenant Junior Grade Noah Hayward, Lieutenant Brent Kinman, Chief Bob Lewis, Petty Officer Third Class Jason Lawson, Petty Officer Second Class Keith Williams, Chief Stacy Hines, and Chief Todd Snyder.

Chapter 12

This chapter is based on interviews conducted aboard the USS *Nebraska* and at the Kings Bay Naval Base with Seaman Ryan Beasley, Lieutenant Brent Kinman, Chief Jeff Spooner, Master Chief Marc Churchwell, Petty Officer First Class Johnnie Jackson, Commander Dave Volonino, Chief Todd Snyder, Petty Officer First Class Don Katherman, Lieutenant Junior Grade Noah Hayward, and Lieutenant Steve Habermas. The author also observed classes in the dive trainer at the Kings Bay Naval Base's Trident Training Facility.

Chapter 13

This chapter is based on interviews conducted aboard the USS *Nebraska* with Lieutenant Stephen Habermas and Petty Officer First Class Rodney Mackey.

Chapter 14

This chapter is based on interviews conducted aboard the USS *Nebraska* and at the Kings Bay Naval Base with Commander Dave Volonino, Lieutenant Fred Freeland, Lieutenant Junior Grade Chad Thorson, Lieutenant Junior Grade Noah Hayward, Lieutenant Commander Harry Ganteaume, Lieutenant Stephen Habermas, Lieutenant Al Brady, Captain Gerald Hunnicutt, Chief Dan Montgomery, Petty Officer First Class Eric Liebrich, Chief of the Boat David Weller, Petty Officer First Class Paul Dichiara, Petty Officer First Class Ed Martin, Petty Officer First Class Mark Lyman, Petty Officer First Class Kevin Jany, Chief Sean Mc-Cue, Petty Officer First Class Jose Victoria, Deanne Victoria, Petty Officer Gregory Murphy, Petty Officer Second Class Keith Williams, and Chief Stacy Hines. Bruce G. Blair also was interviewed.

Also consulted were *The Logic of Accidental Nuclear War*, Bruce G. Blair (Washington, D.C.: Brookings Institution, 1993); and "The Role of American Churches in the Nuclear Weapons Debate," Judith A. Dwyer, in *The Nuclear Freeze Debate: Arms Control Issue for the 1980s*, Paul M. Cole and William J. Taylor, Jr., eds. (Boulder, Colo.: Westview Press, 1983), pp. 77–89. The author also observed training in the missile control center and missile compartment trainer at the Kings Bay Naval Base's Trident Training Facility.

Chapter 15

This chapter is based on interviews conducted aboard the *USS Nebraska* and at the Kings Bay Naval Base with Commander Dave Volonino, Lieutenant Brent Kinman, Ensign Mark Nowalk, Lieutenant Junior Grade Ryan Hardee, and Chief Jeffrey Spooner.

Chapter 16

This chapter is based on interviews conducted aboard the *USS Nebraska* and at the Kings Bay Naval Base with Commander Dave Volonino, Lieutenant Al Brady, Lieutenant Junior Grade Chad Thorson, Lieutenant Brent Kinman, Lieutenant Steve Habermas, Chief Robert Philbin, and Chief Marvin Abercrombie. The author also drew from diaries or letters kept during the patrol by Commander Volonino, Lieutenant Kinman, Lieutenant J. G. Thorson, and Chief Abercrombie.

Chapter 17

This chapter is based on interviews conducted aboard the *USS Nebraska* and at the Kings Bay Naval Base with Lieutenant Ryan Hardee, Lieutenant Brent Kinman, Commander

Dave Volonino, Petty Officer Second Class Matt Suzor, Chief Marvin Abercrombie, Wanda Abercrombie, and Chief Robert Philbin.

The author also drew from diaries or letters kept during the patrol by Commander Volonino, Lieutenant Kinman, and Chief Abercrombie. Also consulted were the June 5, 1999, radio message from the Kings Bay Naval Base on the birth of Sarah Elise Suzor and the June 12, 1999, radio message from the Atlantic Fleet on medical advice for Chief Abercrombie.

Chapter 18

This chapter is based on interviews conducted aboard the USS *Nebraska* and at the Kings Bay Naval Base with Petty Officer First Class Eric Liebrich, Chief Stacy Hines, Chief Marvin Abercrombie, Chief Shawn Brown, Chief Richard McCloud, Petty Officer First Class Mike Frego, Petty Officer Second Class Jason Barrass, Lieutenant Junior Grade Chad Thorson, Lieutenant Brent Kinman, Commander Dave Volonino, Chief of the Boat David Weller, Petty Officer Second Class Chris Wilhoite, and Petty Officer First Class Johnnie Jackson. Also interviewed were Molly McCue, Wanda Abercrombie, Kyung Thorson, Terri Migliore, Vicky Lyman, Jenifer Kinman, and Jenny Brady.

Also consulted were the diaries of Commander Dave Volonino, Lieutenant Junior Grade Chad Thorson, and Chief Marvin Abercrombie; "Nebraska Deployment Information," by Commander Dave Volonino; and "USS Nebraska (SSBN-739)(B)" family gram form.

Chapter 19

This chapter is based on interviews conducted aboard the USS *Nebraska* and at the Kings Bay Naval Base with Commander Dave Volonino, Petty Officer Second Class Chris Wilhoite, Chief Robert Philbin, and Lieutenant Brent Kinman.

Also consulted were the diaries of Commander Volonino and Lieutenant Kinman; "Russian Bombers Make Iceland

Foray," Dana Priest, *Washington Post*, July 1, 1999, p. A1; "Harder Nuclear Arms Policy Is Now Official Kremlin Line," Michael Wines, *New York Times*, Jan. 15, 2000; "Russia Launches Missile," Vladimir Isachenkov, Associated Press report, Dec. 14, 1999; July 5, 1999, radio message from the USS *Nebraska* to the Atlantic Fleet on Petty Officer Wilhoite's medical condition; July 6, 1999; radio message from the Atlantic Fleet to the USS *Nebraska* on Petty Officer Wilhoite's medevac; July 8, 1999, radio message from Submarine Group Ten to the *USS Nebraska* on Petty Officer Wilhoite's medical condition.

Chapter 20
This chapter is based on interviews conducted aboard the USS *Nebraska* and at the Kings Bay Naval Base with Commander Dave Volonino, Lieutenant Brent Kinman, Lieutenant Junior Grade Ryan Hardee, Lieutenant Junior Grade Chad Thorson, and Chief of the Boat David Weller.

Chapter 21
This chapter is based on interviews conducted aboard the USS *Nebraska* and at the Kings Bay Naval Base with Commander Dave Volonino, Chief of the Boat David Weller, Petty Officer First Class Ed Martin, Lieutenant Ryan Hardee, Chief Shawn Brown, Chief Stacy Hines, Petty Officer Second Class Reginald Rose, Chief Rich McCloud, Petty Officer Third Class Frank Levering, Chief Robert Philbin, Seaman Ryan Beasley, Petty Officer Third Class Eric Chambers, Seaman Scott Shafer, Ensign Ray Chesney, Lieutenant Steve Habermas, Lieutenant Junior Grade Rob Hill, Petty Officer First Class Don Katherman, Petty Officer Ashley Fuqua, Ensign Mark Nowalk, Lieutenant Al Brady, and Lieutenant Junior Grade Chad Thorson.

Appendix • The Rest of the USS *Nebraska* Crew

Ryan T. Alfrey
Russell I. Avacato
Jay F. Baker
Brad D. Bishop
Curtis N. Blake
Kelvin D. Blue
Mark K. Bodreaux
Robert A. Bryant
David S. Byford
Christopher L. Catt
Mark C. Chapman
Michael A. Cresho
Christopher M. Davis
Alfredo Donis
Scott W. Doughty
Craig S. Dudley
Terry D. Dunny
James C. Dyer
Brian S. Earls
William F. Engler
Michael J. Frego
Charles C. Gaston, Jr.

Sean S. Gibson
John F. Green
Derek G. Gruell
Andra L. Hall
Brian S. Hallett
Brian M. Haynes
Noah J. Hayward
Bret A. G. Hazeltine
Duane A. Heintzelman
Roger A. Humphrey
Scott A. Humphrey
David A. Huzy
Johnnie L. Jackson, Jr.
John P. Jacobson
Brian D. Johnson
Charles R. Jones, Jr.
Michael P. Jones
Scott A. Kirkland
Robert W. Lewis
Douglas K. Lilley
Steven J. Martin
Scottie L. Martinez

Earl L. Mayes
David L. McClellan
Jeffrey S. Million
Kenneth L. Minatee
Damiso K. Mitchell
Brent M. Newton
Curtis S. Norden
Adam B. Owens
Anthony W. Oxendine, Jr.
Stephen P. Patterson
Adam L. Powers
Maurice E. Powe
Steven E. Randolph
James W. Raycraft, Jr.
Paul M. Rector
William P. Riggs
Levar F. Robinson
Jason D. Rodgers
Scott M. Rua
Kirby J. Schuler
Raymond J. Serbentas

Jeremiah P. Shaw
Gregory W. Smith
Lionel Smith, Jr.
Robert T. Snyder, Jr.
Aaron M. Solomon
Michael R. Steinig
Carl P. Stieren
George W. Stuart III
Richard A. Swartz
Scott A. Thomson
Edward E. Tucker III
John A. Vadnais
Michael D. Valley
Edward P. Verbilla
Garth S. Wall
Nicholas P. Wasley
Michael E. Wheeler
David I. Yates
Christopher A. Zeller
Adam R. Zwiebel

• Index